Springer Proceedings in Mathematics & Statistics

Volume 553

This book series features volumes composed of selected contributions from workshops and conferences in all areas of current research in mathematics and statistics, including data science, operations research and optimization. In addition to an overall evaluation of the interest, scientific quality, and timeliness of each proposal at the hands of the publisher, individual contributions are all refereed to the high quality standards of leading journals in the field. Thus, this series provides the research community with well-edited, authoritative reports on developments in the most exciting areas of mathematical and statistical research today.

Francesco Bartolucci · Paolo Li Donni ·
Fulvia Pennoni · Claudia Pigini
Editors

Models for Longitudinal Data with Applications to Early Warning Systems

Workshops of the PRIN Project, Perugia, Italy, 19 September 2024, 11 September 2025 and 23 January 2026

Springer

Editors
Francesco Bartolucci
University of Perugia
Perugia, Italy

Paolo Li Donni
University of Palermo
Palermo, Italy

Fulvia Pennoni
University of Milano-Bicocca
Milan, Italy

Claudia Pigini
Marche Polytechnic University
Ancona, Italy

ISSN 2194-1009 ISSN 2194-1017 (electronic)
Springer Proceedings in Mathematics & Statistics
ISBN 978-3-032-24140-5 ISBN 978-3-032-24141-2 (eBook)
https://doi.org/10.1007/978-3-032-24141-2

Mathematics Subject Classification: 60G25, 62P20, 62M05, 62M10, 91G45, 62J12, 62P05, 91D30, 62F15

This work was supported by Marche Polytechnic University, Univerisity of Milano-Bicocca, Univerisity of Palermo, University of Perugia.

This Springer imprint is published by the registered company Springer Nature Switzerland AG
The registered company address is: Gewerbestrasse 11, 6330 Cham, Switzerland

Preface

This book is the result of the joint work of a group of researchers who participated in a project funded by the Italian Government (PRIN 2022—Projects of Significant National Interest) entitled “Hidden Markov Models for Early Warning Systems”.[1]

As the title suggests, the project focuses on forecasting rare events by relying on an important class of models for longitudinal data based on unit-specific sequences of latent variables following a Markov chain.

Related models are also considered, together with measures of predictive performance, which are crucial to evaluate the proposed approaches.

Applications span economics and health, as the proposed methods lie at the intersection of Econometrics and Statistics and have a strong connection with Machine Learning (ML).

The project aims to share research ideas and foster collaborations among research groups across the country, in order to increase competitiveness.

This is reflected by the presence of four local research units: at the University of Perugia, the leading unit, which includes the Principal Investigator, Francesco Bartolucci; at Marche Polytechnic University in Ancona, coordinated by Claudia Pigini; at the University of Milano-Bicocca, coordinated by Fulvia Pennoni; and at the University of Palermo, coordinated by Paolo Li Donni.

The preliminary results of the research activity of the project team were presented at two workshops held at the University of Perugia (IT) in September 2024[2] and in September 2025[3].

The final workshop was held at the University of Perugia in January 2026 and provided an opportunity to disseminate the main results achieved by the research team to the general scientific community.

The present book has eight chapters, which are summarized below.

---

[1] For a detailed description, see https://sites.google.com/view/hmm-ews/home

[2] https://sites.google.com/view/hmm-ews/1st-workshop

[3] https://sites.google.com/view/hmm-ews/2nd-workshop

The first chapter investigates the predictive performance of ML-based Early Warning Systems (EWSs) for banking crises, focusing on the application of resampling and cost-sensitive classification methods specifically designed to handle rare events. Using the Systemic Banking Crises Database matched with selected macroeconomic indicators, the chapter shows that these techniques lead to substantial improvements in the F1 score over baseline models. The chapter provides benchmark results that help frame the empirical analysis developed in the subsequent parts of the volume.

The second chapter concerns automated ML techniques for forecasting.

The study shows, in particular, some of these techniques can partially automate the selection and tuning of ML pipelines, thereby reducing manual effort while achieving competitive performance. The results indicate that automated ML provides a valuable starting point for developing financial crisis EWSs, although further manual refinements remain necessary.

The third chapter explores different statistical models in terms of their performance in short-term prediction of binary responses.

These approaches range from logit regression to hidden Markov models and are compared on the basis of country-level longitudinal data, where the response of interest is an indicator variable for the occurrence of a financial crisis. The comparison between models is based on widely used predictive performance metrics.

The fourth chapter illustrates how the regularization due to the adoption of the grouped fixed effects approach can be effectively applied to early warning identification and financial crisis forecasting. Drawing on the Macro History Database, the chapter shows how the approach proves effective in detecting early warnings and forecasting crises, delivering predictions that outperform those obtained with ML methods, while retaining the added benefit of interpretability.

The fifth chapter proposes a Bayesian Student's t-hidden Markov model to capture temporal patterns in financial time series, some of which can be associated with critical market conditions.

For parameter estimation, a data-augmented Markov chain Monte Carlo scheme is adopted, which also allows the dynamic assignment of observations to the latent states.

The methodology is applied to the analysis of the time series of log-returns of certain cryptocurrencies and therefore concerns markets presenting specific peculiarities.

The sixth chapter deals with link prediction in temporal networks, that is, social networks observed over time. Data of this type are modeled using dynamic stochastic block models, in which nodes can move among a set of latent blocks according to a hidden Markov process.

The model is estimated using a variational approach, in which the log-likelihood is suitably approximated and edges are predicted based on the most likely latent group assignments.

The seventh chapter develops a dynamic structural model for longitudinal count outcomes to analyze substitution between primary care and emergency secondary care among individuals with chronic conditions. Building on a hidden Markov bivariate Poisson specification, the model separates unobserved time-varying heterogeneity from state dependence in utilization. Results are mainly summarized as percentage changes in service use, thereby complementing existing policy-evaluation evidence on substitution effects.

Chapter eight exploits a Bayesian bivariate Poisson model to examine how patients use different healthcare services, with particular attention to the balance between primary and secondary care. The analysis focuses on general practitioner consultations and emergency department visits among older adults newly diagnosed with selected conditions, using data from the UK Clinical Practice Research Datalink. The chapter highlights the key drivers of utilization across care settings and discusses their implications for service planning.

We are confident that the content of the book will be of interest to researchers in econometrics, statistics, and machine learning, as well as to practitioners with a particular interest in predicting rare events using longitudinal data.

Perugia, Italy — Francesco Bartolucci
Palermo, Italy — Paolo Li Donni
Milan, Italy — Fulvia Pennoni
Ancona, Italy — Claudia Pigini
January 2026

Acknowledgments This volume and the First, Second, and Final Workshops of the project, held on September 19, 2024, September 11, 2025, and January 23, 2026, were supported by the research project *Hidden Markov Models for Early Warning Systems* (PRIN 2022, project code 2022TZEXKF), funded by the European Union—NextGenerationEU, Mission 4, Component 2, CUP J53D23004990006, under the auspices of the Ministero dell'Università e della Ricerca. We gratefully acknowledge this support.

Contents

Sampling-Based and Cost-Sensitive Classification in Early Warning Systems for Financial Crises

Alex Mircoli, Claudia Pigini, and Domenico Potena

Abstract This study investigates the predictive performance of machine-learning-based early warning systems for banking crises, focusing on methods tailored to the classification of rare events. Using the Systemic Banking Crises Database (1983–2017) matched with macroeconomic indicators, we evaluate the combined impact of *kmeans*-SMOTE oversampling and the cost-sensitive MetaCost algorithm across multiple classifiers. Our findings show that these techniques, especially when applied together, significantly improve the F_1-score over baseline models, with the best configuration achieving a +0.11 gain. Further enhancements are obtained through crisis-based clustering and the inclusion of lagged predictors, which yield the highest predictive performance. Results highlight the complementary benefits of resampling, cost-sensitive learning, and historical information for macro-prudential policy tools aimed at anticipating Systemic Banking Crises.

Keywords Class imbalance · Clustering · Cost-sensitive learning · Sampling

1 Introduction

Early Warning Systems (EWSs) are prominent in several fields of the applied research, as they are employed to forecast events such as natural disasters (Cremen and Galasso, 2020), conflict outbreaks (Cook et al., 2020), and even COVID-19 mortality (Gao et al., 2020). It is in macroeconomics, however, that EWSs as econometric models have received increasing attention from the empirical literature, which also provided their most recent methodological developments.

A. Mircoli (✉) · C. Pigini · D. Potena
Marche Polytechnic University, Ancona, Italy
e-mail: a.mircoli@staff.univpm.it

C. Pigini
e-mail: c.pigini@staff.univpm.it

D. Potena
e-mail: d.potena@staff.univpm.it

F. Bartolucci et al. (eds.), *Models for Longitudinal Data with Applications to Early Warning Systems*, Springer Proceedings in Mathematics & Statistics 553,
https://doi.org/10.1007/978-3-032-24141-2_1

The aftermath of the global financial crisis has spurred the implementation of policy responses aimed at preventing, rather than containing, financial risks. For macro-prudential interventions to be effective, it is therefore crucial to anticipate the generation of financial vulnerabilities at the macro-level. EWSs have therefore attracted increasing attention, in both the policy and academic context (Lang et al., 2018). Policy-oriented studies often build EWSs on panels of countries and specify them as models classifying crisis events based on predictors that can supposedly warn about the occurrence of a crisis several periods in advance.

The first EWSs were built on signal extraction methods, aimed to identify whether a macro-indicator was able to warn about the prominent occurrence of financial distress (Borio et al., 2002; Kaminsky and Reinhart, 1999). Later, logit-based EWSs have been found to outperform country-specific signal extraction EWS as concerns in and out of sample predictions (Davis and Karim, 2008; Filippopoulou et al., 2020). Extensions have dealt with dynamic formulations of the logit/probit model (Alessi and Detken, 2018; Antunes et al., 2018; Caggiano et al., 2016; Pigini, 2021), with the possibility of including country-specific heterogeneity (Dawood et al., 2017; Pigini, 2021; Schularick and Taylor, 2012; Shen and Hsu, 2022), and with their specification as hidden Markov models (see Chap. "Exploring Binary Regression and Hidden Markov Models for Early Warning Systems" of the present book).

The increasing popularity of machine learning methods in the financial field (Nazareth and Reddy, 2023) has spurred the application of novel, more flexible techniques for the classification of financial crises. A first example is provided by Manasse and Roubini (2009), who used classification trees to forecast sovereign debt crises. Then trees and random forests have been widely employed to forecast banking crises (Alessi and Detken, 2018). Also, artificial neural networks play an important role in this literature: after the first contribution by Fioramanti (2008), Ristolainen (2018) analyzed banking crises using country-level data, Tölö (2020) paid attention to systemic financial crises, while Barthélémy et al.(2024) focused on currency crises. Other works focused on bank-level data exploiting different techniques such as support vector machines (Gogas et al., 2018), logistic LASSO (Lang et al., 2018), extreme gradient boosting (Carmona et al., 2019) and their combination (Drudi and Nobili, 2021), and AdaBoost (Casabianca et al., 2022). Finally, machine learning methods lend themselves to the construction of ensemble models to use as EWSs (Liu et al., 2022) and to automation of work pipelines and parameters tuning (see Chap. "Auto Machine Learning for Early Warning Crisis Detection" of the present book).

Several authors compared the performance of standard regression models and machine learning methods. Beutel et al. (2019) did not find a substantial benefit from ML techniques, whereas Holopainen and Sarlin (2017) and Bluwstein et al. (2023) assessed the clear advantage of machine learning algorithms in terms of accuracy of out-of-sample predictions. However, despite its superiority in forecasting accuracy, extant literature has not yet addressed the issue of improving performance by tailoring machine learning methods to the classification of crisis events, which have the peculiarity of being low probability and high impact. The first feature of the data entails the problem generally known in the machine learning literature as class imbalance (Satyasree and Murthy, 2013), usually mitigated by the use of sampling techniques to restore balance of the output variable labels in the training set. The second leads to misclassification costs that are different according to whether a crisis is missed (false negative) or the EWS issues a false alarm (false positive), which could be dealt with by a cost-sensitive determination of probability thresholds for classification.

In this chapter, we revisit the evidence on the performance of the recently employed machine learning methods in forecasting crisis events, combined with techniques suitable to deal with the prediction of rare events. We employ the Systemic Banking Crises Database issued by Laeven and Valencia (2020), which comprises 129 countries from 1983 to 2017 and where only 7% of the outcomes are crises, matched with World Development Indicators (WDI) and International Financial Statistics (IFS) as macroeconomic indicators.

We first look into the application of a sampling technique known as *kmeans*-Synthetic Minority Oversampling Technique (SMOTE, Chawla et al., 2002; Douzas et al., 2018), aimed to balance the training sample in order to address the poor representation of crisis situations. Secondly, we explore the potential of cost-sensitive classification through the MetaCost algorithm (Domingos, 1999), which allows us to assign different costs to different misclassification errors by wrapping a cost-minimizing procedure around any classifier, without requiring parametric assumptions on the loss function.

To the best of our knowledge, this is the first contribution to assess the gain in performance of applying sampling-based and cost-sensitive algorithms to machine learning classifiers for the prediction of financial crises. A partial exception is represented by Bluwstein et al. (2023), who used over- and undersampling in training, without however assessing the actual increase in accuracy that is delivered by the use of such techniques. Also, Sarlin (2013) introduced a loss function parametrically incorporating policymakers' preferences between false alarms and missed crises, which has however been found to be highly sensitive to small perturbations of the preference parameters (Alessi and Detken, 2014).

Our results show that combining *kmeans*-SMOTE with MetaCost consistently outperformed individual strategies, with the best configuration improving the F_1-score by +0.11 over the baseline. Crisis-based clustering provided a further gain in performance, while adding lagged predictors yielded the highest one (+0.16), demonstrating the complementary benefits of resampling, cost-sensitive learning, and historical information in the early warning of banking crises.

Data elaborations have been performed using RapidMiner (RapidMiner GmbH, 2025). The rest of the chapter is organized as follows: Sect. 2 describes the sampling techniques and cost-sensitive classification methods considered in this contribution; Sect. 3 describes the dataset; Sect. 4 presents the experimental framework; Sect. 5 presents the main results, and Sect. 6 draws conclusions.

2 Methodology

In the following, we describe the sampling techniques and cost-sensitive classification methods that we employ to tackle the issues of class imbalance and different misclassification costs in the Systemic Banking Crises Dataset.

2.1 *kmeans-SMOTE*

Standard sampling methods aim to balance the number of observations among classes by randomly replicating observations. In particular, over- and undersampling work by randomly replicating the minority-class or randomly eliminating majority-class observations, respectively. However, naïve random oversampling can lead to overfitting, as it does not provide the model with any new information about the underlying data distribution, while undersampling can discard potentially very useful information (Batista et al., 2004).

One of the methods that has been put forward to overcome this limitation is the SMOTE algorithm (Chawla et al., 2002) that avoids potential overfitting issues by linearly interpolating two existing minority-class observations, thus generating a genuinely new "artificial" observation. More precisely, given a minority instance $\vec{a}$, one of its k-nearest neighbors $\vec{b}$ is selected, and a synthetic sample $\vec{x}$ is created as follows:

$$\vec{x} = \vec{a} + w \cdot (\vec{b} - \vec{a}), \quad \text{with} \quad w \sim U[0, 1],$$

as illustrated in Fig. 1. This approach expands the region of the minority class in the space spanned by the predictors, making the classifier less biased toward the majority class.

While effective at mitigating between-class imbalance, one limitation of the SMOTE algorithm is that it can exacerbate within-class imbalance. In presence of distributional heterogeneity in the minority class, dense regions generate dispro-

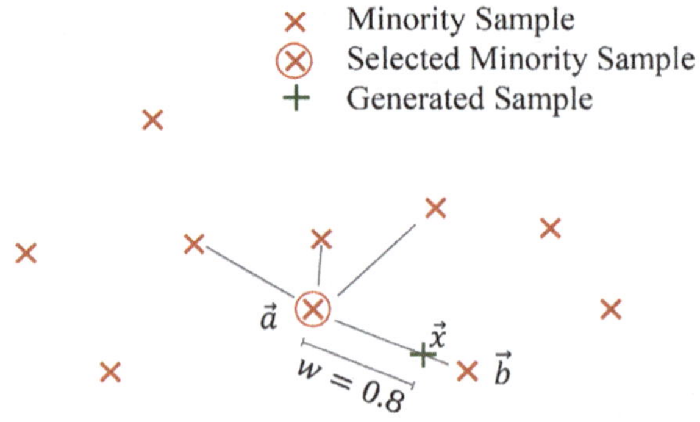

Fig. 1 SMOTE. Source Douzas et al. (2018)

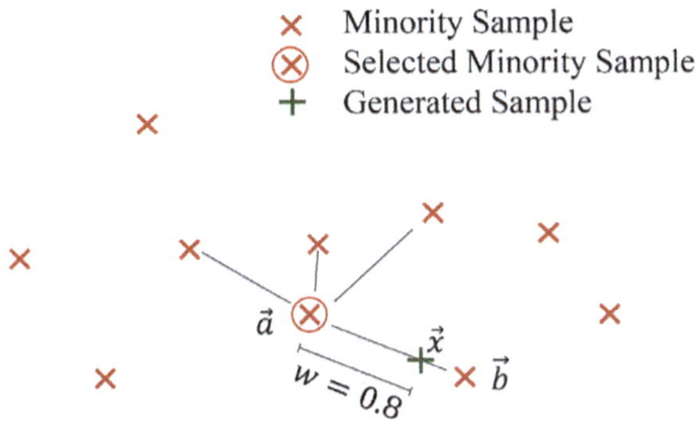

Fig. 2 SMOTE. *Source* Douzas et al. (2018)

portionately more synthetic instances, while potentially informative sparse and/or borderline regions remain underrepresented (see Fig. 2).

The *kmeans*-SMOTE algorithm by Douzas et al. (2018) was developed to address these shortcomings. Their technique introduces a clustering step prior to oversampling. Specifically, minority-class observations are first partitioned into k clusters using the k-means algorithm. The number of synthetic samples to be generated in each cluster is determined based on the local density and the degree of cluster sparsity, favoring less populated regions. The interpolation process, similar to that in SMOTE, is then applied within each cluster (see Fig. 3).

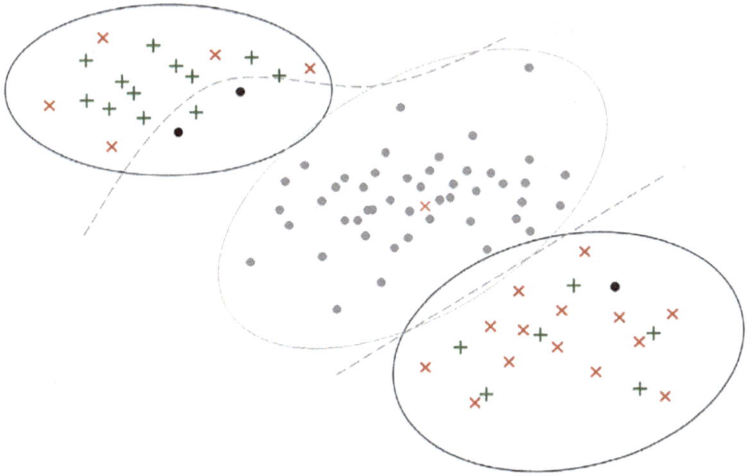

Fig. 3 *kmeans*-SMOTE. *Source* Douzas et al. (2018)

2.2 MetaCost

With financial crises being events characterized by a low probability and high impact, the related cost of misclassifying such events is not uniform: failing to detect a crisis (false negative) may carry far more serious consequences than incorrectly predicting a non-crisis (false positive). Traditional classifiers, however, are typically optimized to minimize the overall error rate, assuming equal costs for all types of errors. Uneven misclassification errors are instead addressed by cost-sensitive classification.

To formalize this idea, define a cost matrix $C(i, j)$, where $C(i, j)$ denotes the cost incurred by predicting an observation as class i when it is actually of class j. In the context of crisis prediction, we typically have $C(1, 0) \ll C(0, 1)$, meaning that misclassifying a non-crisis as a crisis is far less costly than the opposite. Given the posterior class probabilities $\Pr(j|x)$, the Bayes-optimal prediction rule seeks the class assignment that minimizes the expected cost:

$$i^* = \operatorname*{argmin}_i \sum_j \Pr(j|x)C(i, j),$$

which guarantees the lowest expected cost and effectively defines the optimal decision boundary, provided that the class probabilities $\Pr(j|x)$ are known and accurate.

In practice, however, we only have access to estimated class probabilities $\widehat{\Pr}(j|x)$, based on the training dataset. When used in the Bayes-optimal decision rule, these estimates do not necessarily yield the optimal decision boundary, due to sampling variability and model bias. Consequently, even if the classifier is cost-aware, it may still perform sub-optimally.

To address this challenge, MetaCost (Domingos, 1999) is proposed as a general-purpose method for converting any base classifier into a cost-sensitive one. The algorithm works by wrapping a cost-minimization layer around an existing classifier, allowing for an arbitrary cost matrix $C(i, j)$. The core idea is to relabel the training data based on cost-sensitive decisions, in a way such that the cost matrix $C(i, j)$ and estimated class probabilities $\widehat{\Pr}(j|x)$ define the optimal decision boundary.

Specifically, MetaCost estimates the class probabilities $\widehat{\Pr}(j|x)$ through bootstrap aggregation. It generates B bootstrap samples $S_1, \ldots, S_B$, trains the base classifier on each, and averages their predictions to obtain a smoothed probability estimate:

$$\widehat{\Pr}(j|x) = \frac{1}{B} \sum_{b=1}^{B} \widehat{\Pr}(j|x, S_b).$$

Using this ensemble estimate, each training observation is relabeled with the class that minimizes the expected cost:

$$\hat{i} = \operatorname*{argmin}_i \sum_j \widehat{\Pr}(j|x)C(i, j).$$

The classifier is then retrained on this relabeled dataset. By doing so, MetaCost transforms the training data such that the decision boundaries learned by the classifier are approximately cost-optimal.

It is finally worth noticing that the MetaCost algorithm takes care of non-uniform misclassification costs without making use of parametric loss functions. It is in fact in this direction that, to the best of our knowledge, the only attempt at cost-sensitive classification was made in the EWS literature.

3 Data Description

In order to evaluate the performance of machine learning methods combined with sampling-based and cost-sensitive classification, we make use of the Systemic Banking Crises Database issued by Laeven and Valencia (2020). The panel comprises 129 countries, covering the period from 1983 to 2017.

The output variable describes Systemic Banking Crises and it is defined as the occurrence, in a given year, of signs of financial distress in the banking system, such as bank runs, losses in the banking system, and bank liquidations, that have motivated policy interventions as a consequence of significant losses in the banking system.[1] Based on this definition, the sample contains 219 events of systemic banking crises. Table 1 reports the detailed list of crisis episodes identified by Laeven and Valencia (2020).

The list of predictors is compiled following Demirgüç-Kunt and Detragiache (2005), Davis and Karim (2008), and Caggiano et al. (2016). Some descriptive statistics for the dependent and explanatory variables are reported in Table 2 along with their sources. Data on the latter are publicly available as International Financial Statistics (IFS), issued by the International Monetary Fund, or as World Development Indicators (WDI), from the World Bank.

According to a relevant stream of literature on EWSs, early warnings of crisis episodes may be signaled by three main groups of variables. The first one contains fundamental macroeconomic information, such as the real GDP growth rate, the logarithm of per capita GDP, inflation, and the real interest rate. All these variables, with the exception of the last one, are expected to negatively affect the probability of a systemic banking crisis. A second group is identified by monetary variables, specifically broad money (M2) over foreign exchange reserves and the growth rate of private credit. The former represents the (in)ability to face a negative shock to capital inflows and it is therefore expected to positively affect the probability of a crisis occurring. Similarly, the latter is supposed to exert a positive effect on the crisis probability, which can be expected to be increasing with over-indebtedness and deterioration of banks' asset quality. The final group of predictors usually comprises

---

[1] The dataset is available at https://www.imf.org/en/Publications/WP/Issues/2018/09/14/Systemic-Banking-Crises-Revisited-46232.

Table 1 List of crises episodes

Algeria	1990–1994	Argentina	1982, 1991, 1995, 2001–2003
Bangladesh	1987	Belarus	1995
Bolivia	1994	Brazil	1996–1998
Bulgaria	1996–1997	Burundi	1994–1998
Cameroon	1990–1997	Central African Republic	1995–1997
Chad	1992–1996	Chile	1985
China	1998	Colombia	1982, 1998–2000
Costa Rica	1987–1991, 1994	Croatia	1998–1999
Czech Republic	1996–2000	Dominican Republic	2003–2004
El Salvador	1989–1990	Ghana	1982–1983
Guinea	1993	Guinea Bisau	2014–2017
Guyana	1993	Haiti	1994–1998
Hungary	1995–2012	Iceland	2008–2012
India	1993	Indonesia	1997–2001
Jamaica	1996–1998	Japan	1997–2001
Jordan	1989–1991	Kenya	1985, 1992–1994
Korea	1997–1998	Kuwait	1982–1985
Kyrgyz Republic	1996–1999	Lebanon	1990–1993
Malaysia	1997–1999	Mauritania	1984
Mexico	1982–1985, 1994–1996	Moldova	2014–2017
Mongolia	2008–2009	Morocco	1982–1984
Nepal	1988	Nicaragua	1990–1993, 2000–2001
Nigeria	1991–1995, 2009–2012	Panama	1988–1989
Paraguay	1995	Peru	1983
Philippines	1983–1986, 1997–2001	Republic of Congo	1992–1994
Sierra Leone	1990–1994	Swaziland	1995–1999
Sweden	1992–1995	Switzerland	2008–2009
Tanzania	1988	Thailand	1983, 1997–2000
Turkey	2000–2001	Uganda	1994
Ukraine	1998–1999	Ukraine	1998–1999, 2008–2010, 2014–2017
United Kingdom	2007–2011	United States	1988, 2007–2011
Uruguay	1982–1985, 2002–2005	Venezuela	1994–1998
Vietnam	1997	Yemen	1996
Zambia	1995	Zimbabwe	1995–1999

Table 2 Variable sources and descriptive statistics. IFS: International Financial Statistics (International Monetary Fund). WDI: World Development Indicators (World Bank). Inflation is the growth rate of the GDP deflator

	Source	Mean	Median	SD
Dependent variable				
Crisis	Laeven and Valencia (2018)	0.068	0.000	0.252
Explanatory variables				
Real GDP growth (−1)	IFS	3.879	4.089	4.238
Log per capita GDP (−1)	WDI	7.705	7.635	1.458
Inflation (−1)	WDI	26.50	5.940	324.7
Real interest rate (−1)	IFS	2.384	2.125	11.23
M2 to foreign exchange reserves (−1)	WDI	15.21	3.970	33.58
Growth of real domestic credit (−1)	WDI	16.25	12.64	32.86
Growth of net foreign assets to GDP (−1)	WDI	0.019	0.015	0.231

financial information, which is here represented by the growth rate of net foreign assets to GDP and it is expected to negatively affect the likelihood of a systemic banking crisis.

4 Setting

In this section, we describe in detail the comprehensive experimental framework designed to assess the performance of the proposed Early Warning System (EWS) for detecting banking crises. The setup was carefully constructed to ensure methodological rigor, reproducibility, and robustness in the obtained results.

For every machine learning algorithm under consideration, we performed hyperparameter tuning through an exhaustive grid search procedure, ensuring that the search space was sufficiently broad to capture optimal configurations. Model validation was carried out using a tenfold cross-validation scheme to mitigate overfitting and to obtain stable performance estimates.

As primary evaluation metrics, we employed both accuracy and the F_1-score, with particular emphasis on the latter due to the pronounced class imbalance in the dataset. In line with our proposed data preprocessing strategy, we adopted the *kmeans*-SMOTE oversampling technique, combined with undersampling, to mitigate class-imbalance effects while preserving the structure of the minority class.

We evaluated four main classification algorithms:

- Decision Tree (DT): Assessed using multiple splitting criteria (Gini, Accuracy, Information Gain, Gain Ratio), with maximum tree depths ranging from 1 to 135. Both post-pruning and pre-pruning strategies were explored to control model complexity.
- Random Forest (RF): Evaluated under the same splitting criteria as DT, with maximum depths up to 135 and the number of trees ranging from 1 to 400.
- Logistic Regression (LR): Used as a baseline linear classifier for comparative purposes.
- Support Vector Machines (SVM): Tested with linear, polynomial (degrees from 2 to 6), and radial basis function (RBF) kernels. For RBF, the γ parameter took values from the set $[0.0001, 0.001, 0.01, 0.1, 0.2, 0.4, 0.6, 0.8, 1, 10]$.

For each classifier, we systematically evaluated the effect of *kmeans*-SMOTE, the MetaCost algorithm, and their combined application, under various undersampling percentages and clustering strategies. For MetaCost, the cost matrix was treated as a hyperparameter: $C(0, 1)$ varied from 0.05 to 0.5 in steps of 0.05, with $C(1, 0)$ adjusted correspondingly from 0.95 to 0.5. This formulation captures the asymmetric nature of misclassification costs in banking crises, where false negatives (missed crises) are significantly more detrimental than false positives (false alarms). Additionally, we explored the impact of introducing a temporal lag of two periods in the predictors, simulating the effect of delayed information.

The specific experimental configurations tested included the following:

1. Baseline:
   - No *kmeans*-SMOTE and no MetaCost.
2. Combinations of *kmeans*-SMOTE and MetaCost:
   - *kmeans*-SMOTE with different oversampling percentages (15%, 25%).
   - MetaCost alone.
   - *kmeans*-SMOTE (15%, 25%) combined with MetaCost.
3. Country clustering approaches:
   - *kmeans*-SMOTE 15% + MetaCost + Crisis-based Clustering: Countries grouped into three categories according to their average crisis duration.
   - *kmeans*-SMOTE 15% + MetaCost + *kmeans* Clustering: Countries clustered according to macro-financial profile similarity.

4. Lagged predictors:

   – *kmeans*-SMOTE (15%, 25%) + MetaCost with predictors lagged by two periods.

Overall, the above design resulted in a large-scale experiment encompassing 162, 000 distinct runs, whose outcomes are reported and analyzed in Sect. 5.

5 Results

In the following, we present and discuss the main results obtained from the experiment described in Sect. 4. The analysis is organized to highlight the impact of the different strategies for handling class imbalance, both individually and in combination, on the predictive performance of the models.

5.1 kmeans-*SMOTE and MetaCost*

The outcomes of the experiments evaluating the considered class-imbalance handling strategies are summarized in Table 3. For each experimental configuration, we report only the classifier achieving the highest F_1-score, as this metric is more informative than accuracy in the presence of strong imbalance.

For the baseline configuration (i.e., without any rebalancing or cost-sensitive method), the best-performing model is the Random Forest classifier, which reaches an accuracy of 90% but only an F_1-score of 0.58. The substantial gap between these two metrics indicates that the model is heavily biased toward predicting the majority

Table 3 Classification performance of kmeans-SMOTE (kS) and MetaCost compared to baseline

Experimental setup	Best classifier	Accuracy	F1-score
Baseline	Random Forest	90%	0.58
kmeans-SMOTE (kS) (15% undersampling)	Random Forest	91%	0.62
kS (25% undersampling)	Random Forest	90%	0.61
MetaCost	SVM (RBF kernel)	91%	0.66
kS (15%) + MetaCost	**Decision Tree**	**92%**	**0.69**
kS (25%) + MetaCost	SVM (polynomial kernel)	92%	0.68

class, a phenomenon commonly observed in highly imbalanced datasets. In practical terms, the high accuracy is misleading, as it primarily reflects the correct classification of non-crisis periods.

When *kmeans*-SMOTE is applied with a 15% undersampling rate of the majority class, the same classifier's F_1-score improves to 0.62 (+0.04 compared to the baseline). However, increasing the undersampling rate to 25% produces no further gain (F_1 = 0.61), suggesting that discarding additional majority-class observations may remove useful information without providing a proportional benefit in balancing the dataset. These modest improvements indicate that resampling techniques alone may not be sufficient to address the imbalance problem effectively.

In contrast, employing a cost-sensitive learning approach yields more notable gains: wrapping an SVM classifier with the MetaCost algorithm increases the F_1-score to 0.66 (+0.08 relative to the baseline). This improvement highlights the ability of cost-sensitive methods to reshape the decision boundaries in favor of the minority class, directly incorporating the asymmetric misclassification costs into the training process.

The most promising results are achieved through a combined strategy: a Decision Tree trained with *kmeans*-SMOTE (15%) and MetaCost attains the highest F_1-score in the set of experiments, reaching 0.69. Interestingly, increasing the undersampling to 25% in this combined setup again fails to provide additional gains, reinforcing the conclusion that a 15% rate is adequate for the given dataset.

Overall, the +0.11 improvement in F_1-score over the baseline obtained by the best configuration demonstrates a synergistic effect: *kmeans*-SMOTE alleviates class imbalance by increasing the representation of crisis cases, while MetaCost fine-tunes the decision boundaries to account for the higher cost of false negatives. This combination leads to a more balanced and practically relevant predictive performance in the context of early warning for banking crises.

5.2 Country Clustering

Given the substantial variability in the characteristics of individual countries, which may be at different stages of the economic cycle and may have different levels of economic development, it can be challenging to construct a single model capable of accurately predicting crises across all countries. Clustering countries into homogeneous groups based on similarity measures may help algorithms identify more specific and relevant patterns and predictive signals, thereby enhancing overall classification performance. For this reason, we divided countries into clusters using two different techniques, namely, by (i) using the crisis-based clustering technique proposed by Caggiano et al. (2016) and (ii) applying the k-means algorithm. Then, we induced a cluster-specific classification model for each cluster using the best experimental configuration found in Sect. 5.1, i.e., *kmeans*-SMOTE with 15% sampling percentage and MetaCost.

Table 4 Classification performance of the combination of the best model presented in the previous section with country clustering techniques

Experimental setup	Best classifier	Accuracy	F_1-score
kS (15%) + MetaCost	*Decision Tree*	*92%*	*0.69*
kS (15%) + MetaCost + crisis-based clustering	**SVM (RBF kernel)**	**90%**	**0.71**
kS (15%) + MetaCost + clustering based on macro-financial profiles	Decision Tree	91%	0.67

The previous best model (in italic) is reported for comparison purposes

Regarding the first clustering technique (Caggiano et al., 2016), countries were divided into three groups: group A included those countries that, during the period 1984–2017, experienced an average duration of banking crises below the dataset mean of 2.86 years of at least 4 months. Group B included countries whose mean crisis duration deviated from the overall average by no more than 4 months. Finally, group C comprised countries that, on average, experienced a crisis duration longer than the mean crisis duration plus 4 months within the observed time frame. The three resulting clusters differ in terms of cardinality and class distribution. Cluster A consists of 2, 085 instances, of which 2, 036 belong to the "No Crisis" class and 38 to the "Crisis" class, with an overall imbalance of 97.65%. Cluster B contains 382 instances (321 "No Crisis" and 61 "Crisis"), resulting in an imbalance of 84.03%. Lastly, cluster C includes 421 instances, of which 341 labeled as "No Crisis" and 80 as "Crisis," corresponding to an imbalance of 80.99%. For each cluster, we applied the same procedure used in the previous experiments, including a new parameter tuning phase carried out through grid search. Upon completion of the testing phase, a weighted average of the obtained classification metrics was computed in order to allow comparison with the results of the other experiments.

Regarding the k-means clustering, the optimal value for the parameter K was determined through the Elbow method. Specifically, we selected a value of K equal to 4. Given that each country appears multiple times in the dataset, cluster assignment was determined using a majority voting strategy.

The results of the experiments are summarized in Table 4. It can be noticed that the first clustering approach had a moderately positive impact on the classification performance, leading to a +0.02 increase in the F1-score. On the other hand, clustering using k-means led to a slight decrease in the F_1-score (-0.02 w.r.t the previous best model), likely due to the high class imbalance within certain clusters, which negatively affected the performance of the corresponding predictors, lowering the average performance.

Table 5 Classification performance with lagged predictors

Experimental setup	Best classifier	Accuracy	F_1-score
kS (15%) + MetaCost + crisis-based clustering	*SVM (RBF kernel)*	*90%*	*0.71*
kS (15%) + MetaCost + predictors lagged by 2 periods	Random Forest	92%	0.72
kS (25%) + MetaCost + predictors lagged by 2 periods	**Random Forest**	**93%**	**0.74**

The previous best model (in italic) is reported for reference purposes

5.3 Lagged Predictors

In the experiments with lagged predictors, we increased the number of features by considering data from 3 years to predict the target variable 1 year later. Therefore, we used data from years *(t-2)*, *(t-1)*, and *t* to predict year *(t+1)*, so as to predict a crisis by taking into account not just the current year, but also additional information from previous years. However, not all the 2888 samples in the dataset had historical data relating to the previous 2 years. As a consequence, the total number of samples decreased from 2888 to 2464, of which 2313 belonging to class "No Crisis" (previously 2709) and 151 to class "Crisis" (previously 179). Therefore, the results of these experiments are not fully comparable to the previous ones due to the small differences in the considered datasets. Table 5 shows the performance obtained by the best model for each sampling percentage.

By adding temporal lags to the features, higher F_1-scores are achieved compared to the previous models. In fact, the two considered experimental configurations reach F_1-scores of 0.72 and 0.74 (i.e., +0.01 and +0.03 compared to the best previous configuration) and accuracy values of 92% and 93%. Contrary to what was observed in previous experiments, in this case, a higher sampling percentage leads to slightly better results. However, it should be noted that, even though we added much temporal information to the model, the improvements with respect to the other experimental configurations are rather small. This may be due to the fact that the already limited amount of training data has been further slightly reduced, while the number of explanatory variables has tripled, leading to a bigger feature space. The confusion matrix of the best configuration is reported in Table 6.

As expected, the performance for the class *No Crisis* is extremely high, as both precision and recall are close to 1. Anyway, despite the high class imbalance (the ratio between *Crisis* and *No Crisis* is 0.06), it can be seen that 90 crises out of 151 are correctly identified (recall=0.59), thus making the combination of resampling, cost-sensitive classification, and lagged features a promising approach for dealing with the detection of such rare events. For the sake of completeness, it has to be noted that the recall for the *Crisis* class is 0.47, which implies a high number of

Table 6 Confusion matrix of the best model (kS (25%) + MetaCost + predictors lagged by 2 periods)

	Predicted Crisis	Predicted No Crisis	Class recall
Actual Crisis	90	61	**0.59**
Actual No Crisis	102	2211	**0.95**
Class Precision	**0.47**	**0.97**	

false alarms. However, as discussed in Sect. 2.2, this type of error usually has a lower classification cost compared to failing to detect a crisis.

6 Conclusions

This study set out to evaluate whether combining sampling-based and cost-sensitive approaches could improve the predictive performance of machine-learning-based EWSs for banking crises, a task made challenging by the rarity and high impact of such events. Using the Systemic Banking Crises Database matched with macroeconomic indicators, we systematically assessed the impact of *kmeans*-SMOTE oversampling and the MetaCost algorithm across multiple classifiers and experimental configurations.

Our findings provide robust evidence that these techniques, particularly when applied jointly, outperform both baseline models and individual strategies. The best configuration, involving a Decision Tree with *kmeans*-SMOTE (15%) and MetaCost, achieved a +0.11 improvement in F_1-score over the baseline. Furthermore, integrating crisis-based clustering and lagged predictors yielded additional gains, with the highest performance attained when all three elements were combined. These results highlight the complementary nature of resampling, cost-sensitive learning, and temporal information in enhancing predictive accuracy.

From a policy perspective, such methodological enhancements have clear implications: by reducing false negatives without a disproportionate increase in false alarms, these EWSs can provide more reliable signals for macro-prudential intervention. Future research could explore extending these strategies to other types of financial crises, testing their robustness in real-time settings, and incorporating additional sources of heterogeneity, such as institution-level or market-based indicators, to further refine the accuracy and timeliness of crisis prediction.

Acknowledgment We gratefully acknowledge Damian Agachi Menna and Riccardo Rocchi for their valuable collaboration in the development and implementation of the data analyses presented in this volume.

References

Alessi, L., & Detken, C. (2014). On policymakers' loss functions and the evaluation of early warning systems: Comment. *Economics Letters, 124*, 338–340.

Alessi, L., & Detken, C. (2018). Identifying excessive credit growth and leverage. *Journal of Financial Stability, 35*, 215–225.

Antunes, A., Bonfim, D., Monteiro, N., & Rodrigues, P. M. (2018). Forecasting banking crises with dynamic panel probit models. *International Journal of Forecasting, 34*, 249–275.

Barthélémy, S., Gautier, V., & Rondeau, F. (2024). Early warning system for currency crises using long short-term memory and gated recurrent unit neural networks. *Journal of Forecasting, 43*, 1235–1262.

Batista, G. E., Prati, R. C., & Monard, M. C. (2004). A study of the behavior of several methods for balancing machine learning training data. *ACM SIGKDD Explorations Newsletter, 6*, 20–29.

Beutel, J., List, S., & von Schweinitz, G. (2019). Does machine learning help us predict banking crises? *Journal of Financial Stability, 45*, Article 100693.

Bluwstein, K., Buckmann, M., Joseph, A., Kapadia, S., & Şimşek, Ö. (2023). Credit growth, the yield curve and financial crisis prediction: Evidence from a machine learning approach. *Journal of International Economics, 145*, Article 103773.

Borio, C., Lowe, P., et al. (2002). Assessing the risk of banking crises. *BIS Quarterly Review, 7*(1), 43–54.

Caggiano, G., Calice, P., Leonida, L., & Kapetanios, G. (2016). Comparing logit-based early warning systems: Does the duration of systemic banking crises matter? *Journal of Empirical Finance, 37*, 104–116.

Carmona, P., Climent, F., & Momparler, A. (2019). Predicting failure in the us banking sector: An extreme gradient boosting approach. *International Review of Economics & Finance, 61*, 304–323.

Casabianca, E. J., Catalano, M., Forni, L., Giarda, E., & Passeri, S. (2022). A machine learning approach to rank the determinants of banking crises over time and across countries. *Journal of International Money and Finance, 129*, Article 102739.

Chawla, N. V., Bowyer, K. W., Hall, L. O., & Kegelmeyer, W. P. (2002). Smote: synthetic minority over-sampling technique. *Journal of Artificial Intelligence Research, 16*, 321–357.

Cook, S. J., Hays, J. C., & Franzese, R. J. (2020). Fixed effects in rare events data: a penalized maximum likelihood solution. *Political Science Research and Methods, 8*, 92–105.

Cremen, G., & Galasso, C. (2020). Earthquake early warning: Recent advances and perspectives. *Earth-Science Reviews, 205*, Article 103184.

Davis, E. P., & Karim, D. (2008). Comparing early warning systems for banking crises. *Journal of Financial Stability, 4*, 89–120.

Dawood, M., Horsewood, N., & Strobel, F. (2017). Predicting sovereign debt crises: An early warning system approach. *Journal of Financial Stability, 28*, 16–28.

Demirgüç-Kunt, A., & Detragiache, E. (2005). Cross-country empirical studies of systemic bank distress: A survey. *IMF Working Paper: Cross-Country Empirical Studies of Systemic Bank Distress-A Survey* (Vol. 5, pp. 1–33)

Domingos, P. (1999). Metacost: A general method for making classifiers cost-sensitive. *Proceedings of the fifth ACM SIGKDD international conference on Knowledge discovery and data mining* (pp. 155–164)

Douzas, G., Bacao, F., & Last, F. (2018). Improving imbalanced learning through a heuristic oversampling method based on k-means and smote. *Information Sciences, 465*, 1–20.

Drudi, M. L. and S. Nobili (2021). A liquidity risk early warning indicator for Italian banks: a machine learning approach. *Bank of Italy Temi di Discussione (Working Paper) no 1337.*

Filippopoulou, C., Galariotis, E., & Spyrou, S. (2020). An early warning system for predicting systemic banking crises in the eurozone: A logit regression approach. *Journal of Economic Behavior & Organization, 172*, 344–363.

Fioramanti, M. (2008). Predicting sovereign debt crises using artificial neural networks: A comparative approach. *Journal of Financial Stability, 4*, 149–164.

Gao, Y., Cai, G.-Y., Fang, W., Li, H.-Y., Wang, S.-Y., Chen, L., Yu, Y., Liu, D., Xu, S., Cui, P.-F., et al. (2020). Machine learning based early warning system enables accurate mortality risk prediction for covid-19. *Nature Communications, 11*, 5033.
Gogas, P., Papadimitriou, T., & Agrapetidou, A. (2018). Forecasting bank failures and stress testing: A machine learning approach. *International Journal of Forecasting, 34*, 440–455.
Holopainen, M., & Sarlin, P. (2017). Toward robust early-warning models: A horse race, ensembles and model uncertainty. *Quantitative Finance, 17*(12), 1933–1963.
Kaminsky, G. L., & Reinhart, C. M. (1999). The twin crises: the causes of banking and balance-of-payments problems. *American Economic Review, 89*(3), 473–500.
Laeven, L. & Valencia, F. (2018). Systemic banking crises revisited. In *IMF Working Papers 2018/206, International Monetary Fund*
Laeven, L., & Valencia, F. (2020). Systemic banking crises database ii. *IMF Economic Review, 68*, 307–361.
Lang, J. H., Peltonen, T. A., & Sarlin, P. (2018). A framework for early-warning modeling with an application to banks Working paper 2182. ECB.
Liu, L., Chen, C., & Wang, B. (2022). Predicting financial crises with machine learning methods. *Journal of Forecasting, 41*, 871–910.
Manasse, P., & Roubini, N. (2009). "rules of thumb" for sovereign debt crises. *Journal of International Economics, 78*, 192–205.
Nazareth, N., & Reddy, Y. V. R. (2023). Financial applications of machine learning: A literature review. *Expert Systems with Applications,219*, Article 119640.
Pigini, C. (2021). Penalized maximum likelihood estimation of logit-based early warning systems. *International Journal of Forecasting, 37*, 1156–1172.
RapidMiner GmbH (2025). *RapidMiner Studio: Data Science Platform* (11.5 ed.).
Ristolainen, K. (2018). Predicting banking crises with artificial neural networks: the role of nonlinearity and heterogeneity. *The Scandinavian Journal of Economics, 120*, 31–62.
Sarlin, P. (2013). On policymakers' loss functions and the evaluation of early warning systems. *Economics Letters, 119*, 1–7.
Satyasree, K., & Murthy, J. (2013). An exhaustive literature review on class imbalance problem. *Int. J. Emerg. Trends Technol. Comput. Sci, 2*, 109–118.
Schularick, M., & Taylor, A. M. (2012). Credit booms gone bust: monetary policy, leverage cycles, and financial crises, 1870–2008. *American Economic Review, 102*, 1029–1061.
Shen, C.-H., & Hsu, H.-H. (2022). The determinants of asian banking crises–application of the panel threshold logit model. *International Review of Finance, 22*, 248–277.
Tölö, E. (2020). Predicting systemic financial crises with recurrent neural networks. *Journal of Financial Stability,49*, Article 100746.

Auto Machine Learning for Early Warning Crisis Detection

Mirko Cesarini, Luca Brusa, Fulvia Pennoni, and Giorgio Vittadini

Abstract In this work, we investigate the application of Automated Machine Learning (Auto ML) techniques to Early Warning Systems (EWSs) for financial crises at the country level. We explore the end-to-end process of applying supervised learning with Auto ML to financial crisis detection, including data preparation, the treatment of imbalanced datasets without relying on synthetic data, and the construction of representative training and testing sets. This study shows how Auto ML can partially automate the selection and tuning of machine learning pipelines (namely, sequences of computational steps, such as preprocessing and classification), thereby reducing the manual effort required while achieving competitive performance. The results indicate that Auto ML provides a valuable starting point for developing financial crisis EWSs, although further manual refinement and ad hoc adjustments of the identified pipelines remain necessary.

Keywords Financial crises prediction · Imbalanced dataset · Longitudinal data · Stratified downsampling

---

Supplementary Information The online version contains supplementary material available at https://doi.org/10.1007/978-3-032-24141-2_2.

M. Cesarini (✉) · L. Brusa · F. Pennoni · G. Vittadini
University of Milano-Bicocca, Milan, Italy
e-mail: mirko.cesarini@unimib.it

L. Brusa
e-mail: luca.brusa@unimib.it

F. Pennoni
e-mail: fulvia.pennoni@unimib.it

G. Vittadini
e-mail: giorgio.vittadini@unimib.it

F. Bartolucci et al. (eds.), *Models for Longitudinal Data with Applications to Early Warning Systems*, Springer Proceedings in Mathematics & Statistics 553,
https://doi.org/10.1007/978-3-032-24141-2_2

1 Introduction

Early Warning Systems (EWSs) are increasingly incorporating Machine Learning (ML) techniques, whose growing popularity in the financial domain (Nazareth & Reddy, 2023) motivates their use for classification, detection, and prediction of financial crises (see also Chap. "Sampling-Based and Cost-Sensitive Classification in Early Warning Systems for Financial Crises"). However, the application of ML techniques in this context involves several critical issues, e.g., possible class imbalance, where the number of crisis cases is significantly lower than that of non-crises in the available data. It is also important to note that misclassification costs are different according to whether a crisis is missed (false negative) or the EWS issues a false alarm (false positive).

In this chapter, we show an application of Auto Machine Learning (Auto ML) techniques based on supervised learning for EWSs. To the best of our knowledge, such methodology is a novelty for the detection of financial crises. The contribution of this chapter is twofold: on one side, it explores the use of ML and specifically of Auto ML in the context of EWSs, including data preparation, handling of imbalanced data management, and constructing training and testing subsets that accurately represent the original data; on the other side, it demonstrates that many of the tedious tasks involved in tuning ML algorithms can be automated.

This chapter is organized as follows: Sect. 2 presents the state of the art on Auto ML, Sect. 3 describes the dataset used and the preparation steps performed before applying ML algorithms, Sect. 4 shows how Auto ML has been used to find the optimal classification pipeline for financial crisis EWSs, and reports the best pipeline performance. Section 5 draws the conclusions and outlines the future work.

2 State of the Art

In this section, we first survey the ML field, focusing on the Auto ML approaches. Then, we address the issue of imbalanced data management, and we illustrate the Python library which is used to identify the best pipeline for early warning classification.

2.1 Machine Learning

In the following, we propose to use a supervised ML approach, with a binary classifier fitted on a training set to predict upcoming financial crises. Note that a ML algorithm fitted on a specific training set is called a *model* in the ML jargon. Supervised ML is getting very popular in modern data-driven research, enabling predictive modeling across a wide range of scientific domains. Recent advances have focused on improving generalization, interpretability, and scalability of models, particularly through ensemble methods and deep learning architectures (LeCun et al., 2015; Mahalle et al., 2024).

Shallow learning versus deep learning Shallow learning and deep learning represent two broad paradigms within supervised ML. Shallow learning methods, such as logistic regression, support vector machines, and decision trees, rely on handcrafted features and typically involve a single or a few layers of transformation from the covariates to the target variable (Hastie et al., 2009). These models are often computationally parsimonious, and are suitable for problems with small datasets or well-understood feature spaces.

In contrast, deep learning (Nielsen, 2015) employs neural networks with multiple layers, enabling automatic extraction of high-level features from raw data and achieving superior performance in complex tasks such as image recognition, natural language processing, and speech analysis (LeCun et al., 2015; Schmidhuber, 2015). However, deep models generally require large labeled datasets and significant computational resources, and they often lack transparency in decision-making (Lipton, 2018). The choice between shallow and deep learning depends on factors such as data availability, problem complexity, interpretability needs, and computational constraints (Domingos, 2012).

In this chapter, due to the peculiarities of the countries' financial crisis domain, the dataset is too small for applying deep learning approaches. Therefore, only shallow learning classifiers are considered.

Tuning machine learning pipelines ML algorithms often include specific parameters whose optimal values cannot be inferred directly from the training data. In the ML jargon, these are called *hyperparameters*, and identifying their optimal values is crucial for classifier performance. In principle, a brute force approach could be adopted to explore all possible hyperparameter combinations: for each set of values, a classifier is fit on a training set, and its performance is then evaluated on a separate dataset. However, this strategy is rarely feasible, as the search space grows exponentially with the number of hyperparameters. Furthermore, many hyperparameters have an infinite domain, e.g., integer and real numbers. Even when a subset of a parameter domain is chosen,[1] a ML pipeline can feature a lot of hyperparameters, and searching the grid of all possible values may be too computationally expensive. The systematic exploration of hyperparameter combinations is known as *grid search* in the ML jargon. Variants such as random search, which samples from a subset of the grid to reduce computational effort, while commonly used, have shown limitations in terms of efficiency and performance (Yaloveha et al., 2022). In short, the issue can be summarized as follows: the larger the grid to explore, the better the configuration that can be identified, but at the cost of exponentially increasing computation time and resources. To address these challenges, alternative approaches have been developed, including the use of Auto ML as discussed in Sect. 2.2.

[1] Focusing on the C hyperparameter of the `SVM linear` classifier, although its domain ranges on all positive real numbers, frequently the optimal value is searched among a set of powers of 10, e.g., $10^{-3}, 10^{-2}, 10^{-1}, 10^{0}, 10^{1}, 10^{2}, \ldots$.

```
# Library import
from sklearn.pipeline import Pipeline
from sklearn.preprocessing import StandardScaler
from sklearn.linear_model import LogisticRegression

# Pipeline creation
pipeline = Pipeline([          # Define the pipeline
    ('scaler', StandardScaler()),          # Step 1: Standardize
                                           # features
    ('classifier', LogisticRegression()) # Step 2: Logit clsf
])

# Using the pipeline
X_train, X_test, y_train, y_test = ...
# Fit the pipeline on the training data
pipeline.fit(X_train, y_train)
# Make predictions
y_pred = pipeline.predict(X_test)
# ...
```

Listing 1 Example of classification pipeline featuring a standard scaler and a logistic regression classifier

Pipeline In the ML field, pipelines such as the one shown in Listing 1 are frequently used. A pipeline defines a sequence of computational elements, or steps, typically implemented as code modules. Each step processes the input received from the preceding step and produces an output that serves as input for the following step. The first and last elements are exceptions, which respectively process the pipeline's original input data and return the pipeline's output.

In the `scikit-learn` library (Pedregosa et al., 2011), each element in a pipeline has to implement the `fit()` and the `transform()` methods; the only exception is the final element, which is usually a classifier or a regressor, and it must implement the `fit()` and `predict()` methods. The `fit()` method is called when an algorithm is required to learn an internal state from the data that will be used later (e.g., the logit classifier learns once some internal weights during the training, and it will reuse them several times later during prediction); the `transform()` method applies a transformation to the input data (e.g., the `StandardScaler` which standardizes numerical features by removing the mean and scaling to unit variance).

When `fit()` is called on a pipeline, it sequentially calls both the `fit()` method and then the `transform()` method on each element to process the input data, passing the output to the next step. This process continues until the final element (also called *estimator*), where only the `fit()` method is called. During inference, the `predict()` method is called on the pipeline, which calls the `transform()` method on every element of the pipeline except the last one, where the `predict()`

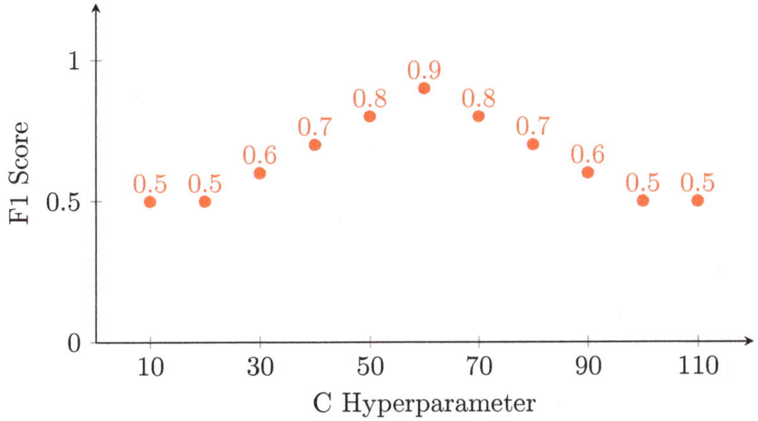

Fig. 1 Hyperparameter values and corresponding F1 score performance plot. Each point represents a specific C value and its corresponding F1 score. It is useful to guide a manual search of the optimal C value that maximizes performance

method is called. This abstraction improves code readability, facilitates reproducibility and parameter tuning. Focusing on the latter, the pipeline hyperparameters can be specified after the pipeline creation, thus facilitating the development of tools that partially or fully automate the hyperparameter tuning process.

2.2 (Auto) Machine Learning

Traditional ML workflows require significant manual effort and domain expertise for tasks such as feature selection, model selection, and hyperparameter tuning. Automated ML has emerged as an approach to automate these phases. The following example illustrates the intuition behind Auto ML: consider the task of exploring values of a hyperparameter C in the range from 1 to 110, to identify the value that maximizes a classifier's performance. At the beginning, 11 randomly picked values for C are considered, and the corresponding F1 scores[2] are shown in Fig. 1 to provide evidence on how hyperparameter variation may affect the performance of a classifier.

Based on the preliminary results of Fig. 1, it would be wise to narrow the search space between 40 and 80. Advanced optimization techniques can be used to reduce the effort of systematically exploring the hyperparameter domain space and to identify the optimal model, e.g., Bayesian optimization (Snoek et al., 2012), evolutionary algorithms (Real et al., 2019), and reinforcement learning (Zoph & Le, 2017). Mod-

[2] The F1 score is the harmonic mean of the precision and recall, with $precision = TP/(TP + FP)$ and $recall = TP/(TP + FN)$. TP, FP, TN, and FN are, respectively, the true positives, false positives, true negatives, and false negatives. It summarizes precision and recall, and it is not affected when precision is improved at the expense of recall or vice versa.

ern Auto ML pipelines typically evaluate components for automatic data cleaning, transformation, model evaluation, and ensemble construction (Feurer et al., 2015).

Although a detailed comparison of the different Auto ML approaches is outside the scope of this chapter, we briefly mention that several open-source and commercial platforms have significantly advanced the accessibility of these techniques. These include Auto-sklearn (Feurer et al., 2015), which integrates meta-learning and ensemble selection, Tree-based Pipeline Optimization Tool (TPOT) (Olson & Moore, 2016), which uses genetic programming for pipeline optimization, AutoKeras (Jin et al., 2019) for deep learning tasks, and cloud-based services such as Google Auto ML (Google, 2025) and Microsoft Azure Auto ML (Azure, 2025).

Recent research has focused on developing more efficient, adaptive Auto ML frameworks. Emerging trends include few-shot and zero-shot Auto ML[3] using transfer and meta-learning (He et al., 2021), multi-objective optimization that balances accuracy with latency or interpretability (He et al., 2021), and human-in-the-loop systems that integrate expert feedback to guide search processes more effectively.

Here, we propose the use of `TPOT` as the Auto ML tool since it explores and optimizes ML pipelines using genetic programming (Olson & Moore, 2016). `TPOT` provides a predefined set of pipeline components and corresponding hyperparameter search spaces, thereby streamlining the model selection and tuning process without extensive manual intervention (Olson et al., 2016). Its use of genetic programming enables efficient exploration of the search space, often yielding high-performing models with fewer iterations compared to traditional grid or random search approaches (Azedou et al., 2025; Leung et al., 2003). Additionally, `TPOT` outputs a fully executable Python script (Python Software Foundation, 2025; van Rossum & Drake, 2011) that reproduces the selected pipeline and hyperparameter configuration, enhancing transparency and reproducibility. It also reduces the work of creating and testing the identified pipelines. Importantly, `TPOT` pipelines are constructed exclusively from components of the `scikit-learn` library, ensuring compatibility with widely used ML standards and facilitating integration into data processing workflows (Pedregosa et al., 2011). Further information about `TPOT` is provided in Sect. 2.4.

2.3 Imbalanced Data Management

In supervised ML, a dataset is considered imbalanced when instances of one class significantly outnumber those of the other one(s), which poses a challenge to model performance. A classifier trained on an imbalanced dataset tends to be biased toward

[3] *Zero* (or *few*) *shot classification* refers to the task of predicting a class that was not observed during training (or where only a limited number of examples were provided). This method often involves the use of transfer learning, i.e., a model trained on a different task is adapted for the target task. The terms few-shot and zero-shot classification originated in natural language processing, where large language models and prompting techniques were employed to build zero-shot classifiers (Perez et al., 2021).

the majority class, leading to poor generalization and low predictive performance for the minority class. This issue is particularly critical in domains where the minority class carries high importance, such as fraud detection, medical diagnosis, and anomaly detection (He & Garcia, 2009). Traditional evaluation metrics like accuracy can be misleading in these contexts, as a model may achieve high accuracy simply by predicting the majority class (Chawla et al., 2002). Several techniques exist to address or mitigate the imbalanced data problem, including down-sampling, cost-sensitive learning, and ensemble methods that focus on minority class detection (Krawczyk, 2016).

EWSs based on financial data, like those considered in the present work, are highly imbalanced, since the crisis cases are few compared to the non-crises. Therefore, we focus on down-sampling the majority class by replacing its records with a randomly selected subset, ensuring the new subgroup has the same dimension as the crisis set. In this way, we explore the predictive performance of the ML algorithms on EWSs without adding synthetic data. Ad hoc stratification is used to preserve the data balance between the extracted subset and the original one, as illustrated in Sect. 3.1.

2.4 Tree-Based Pipeline Optimization Tool

`TPOT` is an open-source Python library that leverages genetic programming to identify and optimize ML pipelines (Olson & Moore, 2016). Built on top of `scikit-learn`,[4] `TPOT` automates model selection, feature preprocessing, and hyperparameter tuning using a genetic approach to optimize pipelines.

In `TPOT`, a pipeline is represented as a tree,[5] with pipeline elements as nodes, and a set of predefined pipelines is referred to as a *population*. An example of a tree representing a pipeline is shown in Fig. 2.

The genetic algorithm used by `TPOT` is inspired by Charles Darwin's theory of evolution (survival of the fittest), in which the fittest individuals have a higher probability of being selected for reproduction. Pipeline elements evolve through operations such as mutation and crossover. The population is evaluated, and only the best-performing elements are passed to the next generation (Mitchell, 1998). `TPOT` creates the initial population by randomly generating a fixed number of trees, each tree representing a ML pipeline, with elements randomly drawn from a list of classifiers, feature selectors, and other components. `TPOT` provides some predefined lists (e.g., "TPOT Light", which restricts the model selection to the shallow learning classifiers available in the `scikit-learn` library) and also allows users to create custom lists. For further details, the interested reader can refer to the *TPOT documentation*

[4] A Python library widely used in the community (Pedregosa et al., 2011).

[5] Although each tree node can have only one parent node, this constraint is not respected in `TPOT` pipelines. Nevertheless, by abuse of notation, `TPOT` pipelines will be called trees in this work.

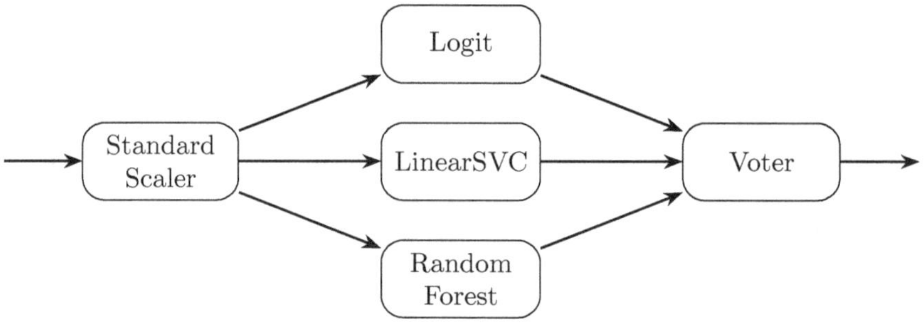

Fig. 2 Example of a pipeline implementing an ensemble classifier: the pipeline starts with a `standard scaler` followed by three classifiers (`logit`, `linearSVC`, and `random forest`) whose predictions are combined using a `Voter`. The voter determines the final output based on the majority of the classifier predictions

(EpistasisLab, 2025). Focusing on the next generation, new trees (offspring) are created in a manner analogous to the formation of new chromosomes:

- either by merging two trees from the current generation using the crossover operation (starting from two randomly selected pipelines, both are split at a random point and their contents are swapped);
- or by modifying an element using a mutation operator (inserting, deleting, or randomly replacing an element in the tree).

The new offspring pipelines are checked, and invalid ones are dropped; e.g., a pipeline that attempts to pass a feature set into an element that expects a single integer would be dropped. This can be easily checked because, during the performance evaluation (described next), a malformed pipeline will raise a computation error. Each tree (both the old and the offspring ones) is evaluated using a fitness function, e.g., the classification accuracy (or any available machine learning metric, e.g., F1 score, AUC-ROC, ...) and the pipeline complexity (pipelines using few elements are preferred). Finally, using the evaluation results, a new generation is formed by selecting the fittest pipelines, ensuring that the population size remains constant. Hyperparameters are managed in a similar way to nodes of the tree. Throughout the process, the single best-performing pipeline discovered during the `TPOT` run is stored separately and serves as the representative pipeline afterward. Through genetic evolution, `TPOT` can efficiently discover high-performing pipelines that might not be apparent through manual design.

Unfortunately, optimization approaches based on genetic algorithms suffer from limitations such as getting stuck in local minima. Enlarging the available dataset or lengthening the number of generations could mitigate or address the problem. Despite these limitations, Auto ML tools like `TPOT` are valuable for practitioners seeking effective models without extensive manual tuning. Of course, the pipeline identified by `TPOT` is not the final step, but often marks the beginning of further research efforts; an example is illustrated in Sect. 4.2.

3 Dataset and Preliminary Preparation

The data considered for the applicative example are the same as those illustrated in Chap. "Sampling-Based and Cost-Sensitive Classification in Early Warning Systems for Financial Crises", also used in Chaps. "Exploring Binary Regression and Hidden Markov Models for Early Warning Systems" and "A Regularized EWS for Banking Crises: A Grouped Fixed Effects Approach", and originally presented in Pigini (2021). For the current proposal, they require some adjustments to achieve the best predictive performance with the ML algorithms. In the following, after briefly describing the data, we show how missing values were handled and how additional grouping variables were created to perform stratified downsampling that approximates the original dataset composition as closely as possible. Grouping variables have been built using both data from the original dataset and information from external sources (e.g., country dimensions, geographical location, etc.).

3.1 *Data Description*

The data used for the application comprise a binary target variable, named `Crisis`, indicating whether a systemic banking crisis occurred across 129 countries over 35 years, from 1983 to 2017. As illustrated in Chap. "Sampling-Based and Cost-Sensitive Classification in Early Warning Systems for Financial Crises", a total of 219 crisis events are observed, compared to 2,986 non-crisis instances. As expected, non-crisis episodes significantly outnumber crisis occurrences. A set of macroeconomic, monetary, and financial indicators, each lagged one year relative to the target variable, are included as covariates. These comprise the `Real GDP Growth Rate`, the `Logarithm of the per Capita GDP`, the `Inflation Rate`, the `Real Interest Rate`, the `Ratio of Broad Money to Foreign Exchange Reserves`, the `Growth Rate of Private Credit`, and the `Growth Rate of net Foreign Assets to GDP`. The latter variable, in particular, contains 160 missing values. These were imputed using the following procedure: missing values were replaced with 0, and a corresponding dummy variable was created (named `Aux`), taking the value 1 for imputed records and 0 otherwise. In addition, we also include the one-year lagged target variable as a covariate.

3.2 *Grouping Variables*

As previously mentioned, the dataset is highly imbalanced, with a number of crises observed over the considered time span being considerably lower than the number of non-crises. Furthermore, some countries did not experience any crises during the observation period. In this context, using undersampling could result in removing all available non-crisis data for those countries. This issue can be addressed by grouping countries based on their similarities and ensuring that at least 4 elements (2 crisis and 2 non-crisis records) remain in each group after undersampling. Several grouping strategies have been evaluated; the *acceptance criterion* is described as follows: after applying undersampling and train-test splitting, the two resulting subsets should contain at least one crisis record and one non-crisis record for each group of countries. Grouping was based on the following variables:

- Country geographical information: a variable named `Continent` was created with the following categories: North America and Oceania, South America, Europe, Asia, and Africa. Since the Oceania countries had no crises, this group was merged with the North America group because they share a common language and cultural heritage. In addition, the country's land area plus water bodies (lakes, reservoirs, and rivers), measured in square kilometers (km^2) was considered through a variable named `Land Area Km`^{2}.
- The `Logarithm of the per Capita GDP`, previously introduced, which focuses on the economic status of the country.

The rationale for focusing on the three variables mentioned above is to group countries based on geographical and economic similarities. Both variables `Land Area Km`^{2} and `Logarithm of the per Capita GDP` were binned, using the `pandas qcut` function (Qcut, 2025), which discretizes a variable into equal-sized buckets based on quantile cuts computed from the actual values of the variable. Several quantile values were evaluated; the only combination that satisfied the acceptance criterion introduced earlier was two bins per variable. Two additional variables, namely `QR Land Area Km`^{2} and `QR Log of the per Capita GDP`, were created to store the labels produced by the binning process. These variables were not fed to the ML algorithms; they were used only for downsampling and train-test splitting, as outlined in Sect. 3.3.

3.3 Downsampling and Train-Test Split

A variable named `Sel` was added to the dataset, whose values were computed by concatenating the values of the variables `Continent`, `QR Log of the per Capita GDP`, and `QR Land Area Km`^{2}. The `Sel` variable was created to support the downsampling and train-test splitting and can be considered as a stratification variable. An example is shown in Table 1. It is worth noting that the records for *Chile* appear in two different `Sel` groups in Table 1. This occurs because the values of the variable `Logarithm of the per Capita GDP`, depending on the year, fall into both bin 0 and bin 1 of the `QR Log of the per Capita GDP` variable.

The dataset was then balanced within each level of the stratification variable `Sel` by undersampling the majority class (non-crises) to match the minority class size (i.e., crises). The resulting records were subsequently split into training and test sets,

Table 1 Example of the `Sel` variable and the other variables used to build it. The `Country Name` was not used to build `Sel`, but it is reported for information purposes only. The records shown are only a subset of the original dataset

Country Name	Continent	QR Log of the per Capita GDP	QR Land Area Km^2	Sel
Burundi	Africa	0	0	Africa.0.0
Cameroon	Africa	1	0	Africa.1.0
South Africa	Africa	1	1	Africa.1.1
Switzerland	Europe	1	0	Europe.1.0
United Kingdom	Europe	1	1	Europe.1.1
Belarus	Europe	0	1	Europe.0.1
Costa Rica	North America and Oceania	1	0	North America and Oceania.1.0
United States	North America and Oceania	1	1	North America and Oceania.1.1
Argentina	South America	0	1	South America.0.1
Chile	South America	0	0	South America.0.0
Chile	South America	1	0	South America.1.0
...	...	...	...	...

while the discarded observations were retained as additional test subsets for potential future use. In this way, the `Sel` variable was used for performing both undersampling and train-test split as outlined in Algorithm 1 in Appendix A. The algorithm labels each record using one of the following:

- Balanced_Train: label used for records in the balanced train subset;
- Balanced_Test: label used for records in the balanced test subset;
- Test2: label used for records not selected for inclusion in the balanced train and test subsets;
- Test3: label used for the groups where either the crisis or the non-crisis subset was absent.

Table 2 shows the number of elements in each group identified by the labels described above, stratified by crisis and non-crisis.

Table 2 Number of records in the groups Balanced_Train, Balanced_Test, Test2, and Test3. The records dropped during downsampling are all related to the non-crisis class, which is the majority class. All the Test3 elements actually refer to non-crisis events too

Group label	QT crisis	QT Non-crisis
Balanced_Train	156	156
Balanced_Test	63	63
Test2	–	2603
Test3	–	164

3.4 Rolling Window Dataset

The dataset composed of records from the Balanced_Train and Balanced_Test described in Sect. 3.3 was used for training and evaluating the ML algorithm, respectively. For reference, this dataset is referred to as ML_Dataset. Each record contains information for a single country, with the response variable corresponding to year t and all covariates corresponding to year $t-1$. Additionally, we considered a dataset implemented using a five-year rolling window strategy, referred to as ML_Rolling_Dataset hereafter. In this dataset, each record corresponds to a specific country, at year t for the response variables, while the covariates span over the five preceding years $t-1, \ldots, t-5$. When the reference year was too close to year 1983, the earliest year in the original data, there were not enough years to fill the rolling window. Therefore, the last available year was repeated until the rolling window reached the desired size.

Missing combinations of country and year were skipped, with an additional previous year included to maintain the window length. It is worth noting that both the ML_Dataset and the ML_Rolling_Dataset contain the same number of records, each focusing on a specific combination of country and year. The target variable for the optimal pipeline search in both datasets was the feature `Crisis`.

4 Search for the Optimal Classification Pipeline(s)

The search for the optimal pipeline was performed using both the considered datasets (i.e., ML_Dataset and ML_Rolling_Dataset, introduced in Sect. 3.3), and two pools of elements upon which the `TPOT` library builds the initial pipeline population and starts the search (described in Sect. 4.1). The optimal pipelines found are presented in Sect. 4.2. Further analyses on the best-performing pipeline are reported in Sect. 4.3. Section 4.4 compares the proposed pipelines against baseline models from the literature, while Sect. 4.5 examines feature importance through SHAP values. Some considerations are drawn in Sect. 4.6.

4.1 TPOT Pipeline Search Setup

As previously introduced, `TPOT` builds the initial population of pipelines by randomly combining the elements drawn from a pool. Two pools of elements were considered:

- Light_Pool: This is a pool provided by the `TPOT` library, which restricts the model selection to the shallow learning classifiers available in the Python library `scikit-learn`. All other elements were unrestricted; e.g., the feature selection elements were all those available from the `scikit-learn` library.
- Custom_Pool: This is a custom pool in which the set of classifiers and feature extractors is restricted to the following list. According to the authors' experience, these elements are the most frequently used for tabular data classification. The selected elements, along with their respective Python libraries, are listed below; a note is provided whenever the library element name is not self-explanatory:
  - `sklearn.tree.DecisionTreeClassifier`;
  - `sklearn.naive_bayes.BernoulliNB`: Naive Bayes classifier for Bernoulli distributions;
  - `sklearn.ensemble.ExtraTreesClassifier`: implements a meta estimator that fits several randomized decision trees (also known as extra-trees) on various sub-samples of the dataset and uses averaging to improve the predictive accuracy and control over-fitting;
  - `xgboost.XGBClassifier`: eXtreme gradient boosting, is an ensemble learning method that combines multiple weak learners (like decision trees) to create a strong predictive model;
  - `sklearn.linear_model.SGDClassifier`: implements regularized linear models with stochastic gradient descent (SGD) learning;
  - `sklearn.svm.LinearSVC`: linear support vector classification;
  - `sklearn.feature_selection.RFE`: feature ranking with recursive feature elimination;
  - `sklearn.feature_selection.SelectFromModel`: a component that selects features based on importance weights learned by an underlying model.

Moreover, each `TPOT` pipeline search session was started using the following parameters:

- `population_size=100`: the size of the initial population where the genetic optimization is applied;
- `generations=100`: the number of generations; at each generation, new offspring pipelines are created while the less performing pipelines are dropped;
- `random_state=42`: most of `TPOT` operations are based on random initialization of values. This parameter is used as a seed to make random initialization repeatable;

```
from tpot import TPOTClassifier

# ...

xTrain=... # The training set covariates
yTrain=... # The training set labels

tpot = TPOTClassifier(
        generations=100, #Number of iterations
        population_size=100,
        scoring='f1_macro', # the F1 score macro averaged will be used to evaluate the pipeline performance
        cv=5, # cross-validation
        n_jobs=-1, # Use all but one available processors in parallel to expedite computation
        config_dict='tpot light',
        random_state=42)

# The next command starts the optimal pipeline search
tpot.fit(xTrain, yTrain)

# The optimal pipeline is saved in a Python script
tpot.export('script.py') # From this script the best pipeline can be extracted
```

Listing 2 Example of code to perform a TPOT optimal pipeline search.

- `cv=5`: when a pipeline is evaluated, 5-folds cross-validation is used on the train set.

An example of `TPOT` invocation is shown in Listing 2. Once the `TPOT` pipeline optimization ends, the resulting pipeline is trained from scratch on the train set and evaluated on the test set, as introduced in Sect. 3.3.

An optimal pipeline search takes about 8 h on an Intel I7 Linux machine with 32 GB of RAM.

4.2 TPOT Pipeline Search Results

For each initial search setup, an optimal pipeline was selected, namely P1.C.M, P2.L.M, P3.C.R, and P4.L.R. The suffixes .x.y in P are the following:

- .x refers to the pool of elements upon which the initial pipeline population is built. Possible values are

Table 3 Optimal pipelines selected for each initial search setup, defined by the combination of element pool and dataset, along with the corresponding F1 score

Pipeline nickname	Element pool	Dataset	F1 score
P1.C.M	Custom_Pool	ML_Dataset	0.79
P2.L.M	Light_Pool	ML_Dataset	0.79
P3.C.R	Custom_Pool	ML_Rolling_Dataset	**0.86**
P4.L.R	Light_Pool	ML_Rolling_Dataset	0.85

The best value of F1 Score is in bold

  - **L**, denoting the Light_Pool;
  - **C**, denoting the Custom_Pool;

- .y refers to the dataset used during the `TPOT` search. Possible values are:

  - **M**, denoting the ML_Dataset;
  - **R**, denoting the ML_Rolling_Dataset.

The initial search setup of each selected pipeline is shown in Table 3, as well as the obtained F1 score, while the pipeline description and the performance on the test set are shown next.

Pipeline P1.C.M The pipeline shown in Listing 3 is an evolution of the best pipeline that `TPOT` identified for the Custom_Pool of elements and ML_Dataset. Sometimes the pipelines can be further improved by applying some changes that a human brain can quickly identify. The original pipeline that `TPOT` identified is reported in Listing 5 and the description of the manual intervention is outlined in Sect. 5.

In Listing 3, the `make_pipeline` class creates a pipeline instance. Its parameters are the pipeline elements, in the exact order they will occupy the pipeline. The first element is the `StackingEstimator`, a class provided by `TPOT` for creating pipelines that combine multiple classifiers. It is a meta-transformer that wraps a classifier and produces an output composed of the input dataset enriched by the classifier predictions (and/or class probabilities). When the `StackingEstimator` is created, a classifier is specified and stored internally. The `StackingEstimator` implements the `fit()`, `transform()`, and `predict()` methods required for pipelines, as outlined in Sect. 2.1. During prediction, the internal classifier processes the input features; its predictions are merged with the original input, and the combined result is passed to the next pipeline element. In this case, several classifier outputs are added to the input dataset, namely the `BernoulliNB` and the `DecisionTreeClassifier`.

Finally, the enriched features are given to the `ExtraTreesClassifier`.

The pipeline classification report is summarized in Table 4.

Despite the manual intervention, both pipelines (P1.C.M.bis and P1.C.M) have similar performance. Therefore, they will be both referred to as P1.C.M, unless otherwise specified.

```
# ...
pipe=make_pipeline(
        StackingEstimator(
            estimator=make_pipeline(
                StackingEstimator(
                    estimator=BernoulliNB(alpha
                        =10.0, fit_prior=True)),
                DecisionTreeClassifier(criterion="
                    gini", max_depth=6,
                    min_samples_leaf=11,
                    min_samples_split=11))),
        ExtraTreesClassifier(bootstrap=False,
            criterion="gini", max_features=0.05,
            min_samples_leaf=2, min_samples_split
            =13, n_estimators=100))
```

Listing 3 The P1.C.M.Bis Pipeline Code

Table 4 Classification report of the pipeline P1.C.M.Bis evaluated on the test set

	Precision	Recall	F1 score	Support
Non-crises (0)	0.76	0.83	0.79	63
Crises (1)	0.81	0.75	0.78	63
Accuracy	–	–	0.79	126
Average	0.79	0.79	0.79	126

Table 5 Classification report of the pipeline P3.C.R evaluated on the test set

	Precision	Recall	F1 score	Support
Non-crises (0)	0.80	0.95	0.87	63
Crises (1)	0.94	0.76	0.84	63
Accuracy	–	–	0.86	126
Average	0.87	0.86	0.86	126

Pipeline P3.C.R The P3.C.R, the best pipeline that `TPOT` identified working on the Custom_Pool of elements and on ML_Rolling_Dataset, is reported in Listing 4. Most of the elements used in Listing 4 have already been introduced. The comments should clarify the new pipeline elements. The classification report is shown in Table 5.

```
from sklearn.pipeline import make_pipeline
from sklearn.naive_bayes import BernoulliNB #Naive Bayes classifier for multivariate Bernoulli models
from sklearn.svm import LinearSVC # Linear SVM Classifier
from xgboost import XGBClassifier
from tpot.builtins import StackingEstimator
pipe=make_pipeline(
        # 1st pipeline element
        StackingEstimator(
            estimator=LinearSVC(
                C=1.0, dual=True, loss="squared_hinge",
                penalty="l2", tol=0.0001)
                # Linear SVM classifier),
        # 2nd pipeline element
        StackingEstimator(
            estimator=BernoulliNB(alpha=10.0, fit_prior=True)),
        # 3rd pipeline element
        XGBClassifier(
            learning_rate=0.01, max_depth=3, min_child_weight=12, n_estimators=100, n_jobs=1, subsample=0.9500000000000001,
            verbosity=0))
```

Listing 4 The P3.C.R Pipeline Code

4.3 *Further Analysis on the P3.C.R Pipeline*

In this section, we focus on the P3.C.R pipeline described in Sect. 4.2, which is the best-performing pipeline according to the results shown in Table 3.

Training curve To investigate possible overfitting, we randomly extract subsets of the training data, considering sizes ranging from 1 to 100% of the training set. For each subset, we train the pipeline on it, then evaluate on both the test set and the training set, and compute the F1 score in both cases. Plotting these two performances against the subset sizes yields the training curves, shown in Fig. 3. The plot in Fig. 3 shows that a training size of 100 elements is sufficient to achieve peak performance. Moreover, enlarging the dataset does not appear to lead to overfitting, as test performance does not deteriorate significantly.

Sensitivity analysis To assess the stability of the results, we also conducted a sensitivity analysis of the P3.C.R pipeline. The pipeline was trained 100 times on the same dataset (the training subset of ML_Rolling_Dataset), and in each run, performance was evaluated on the test set. The mean and standard deviation of the F1 scores were 0.8202 and 0.0015, respectively.

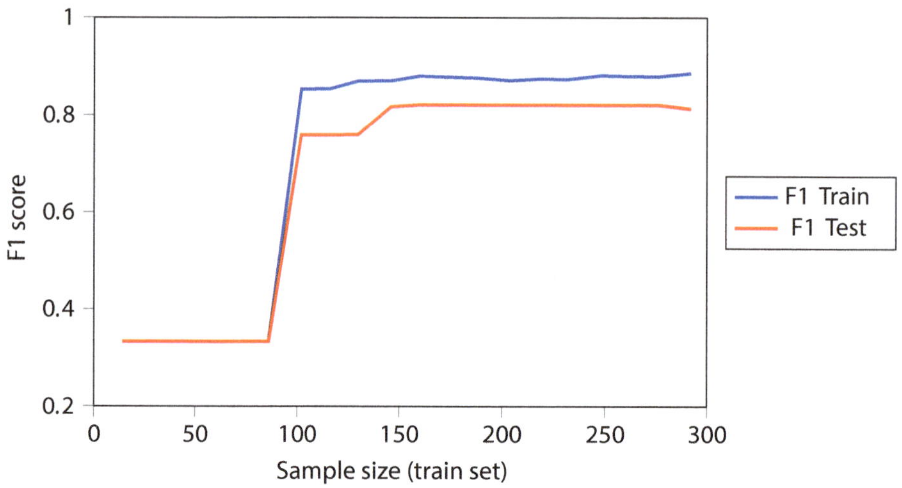

Fig. 3 Pipeline P3.C.R training curve. The x-axis focuses on the train set sample size, while the y-axis focuses on the F1 score performance

4.4 *Comparisons*

The P3.C.R and P1.C.M pipelines described in Sects. 4.2 and 4.3 respectively have been compared with existing models proposed in the literature. The selected models and the rationale for their choice are outlined below.

The `penalized Logit` introduced in Pigini (2021) is among the state-of-the-art models for EWS. Since the dataset used in this chapter is balanced, and the `penalized Logit` is further explored in other chapters of this book, a simple `Logistic Regression` has been used for comparison with the proposed pipelines. According to Pigini (2021), Casabianca et al. (2022) and Li (2025), economists have increasingly employed supervised ML methods, including Random Forests (RF) and XGBoost. Therefore, the following models have been selected as baselines.

- A `Logit` Classifier (McCullagh & Nelder, 2019);
- A `Random Forest` model (Breiman, 2001);
- An `XGBoost` model (Chen & Guestrin, 2016).

A grid search has been performed to identify the best hyperparameters for each baseline model. The P3.C.R and P1.C.M pipelines were excluded from the grid search, since they are already optimized. The base model optimization was performed twice, both on the ML_Dataset and on the ML_Rolling_Dataset. The best hyperparameters were searched on the Balanced_Train subsets of both datasets using a fivefold cross-validation. The best hyperparameters found are shown in Listings 8, 9, and 10 in Appendix C, while Table 6 reports the pipeline performance. Each pipeline was fitted on the Balanced_Train and evaluated on the Balanced_Test. The

Table 6 F1 score, precision, recall and area under the PR curve for both considered datasets and for each model pipeline; precision and recall values are averaged over the two classes (crisis, non-crisis)

Dataset	Pipeline	F1 score	Precision	Recall	PR_AUC
ML_Dataset	Logit	0.776	0.778	0.778	0.806
ML_Dataset	RF	0.777	0.778	0.778	**0.874**
ML_Dataset	XGB	0.769	0.770	0.770	0.862
ML_Dataset	P1.C.M	**0.785**	**0.786**	**0.786**	0.854
ML_Rolling_DS	Logit	0.817	0.817	0.817	0.871
ML_Rolling_DS	RF	0.794	0.794	0.794	0.889
ML_Rolling_DS	XGB	0.848	0.849	0.849	0.890
ML_Rolling_DS	P3.C.R	**0.856**	**0.857**	**0.857**	**0.892**

ML_Rolling_DS is the short for ML_Rolling_Dataset. For each column, the best value for the ML_Dataset and for the Rolling_DS dataset is written in bold.

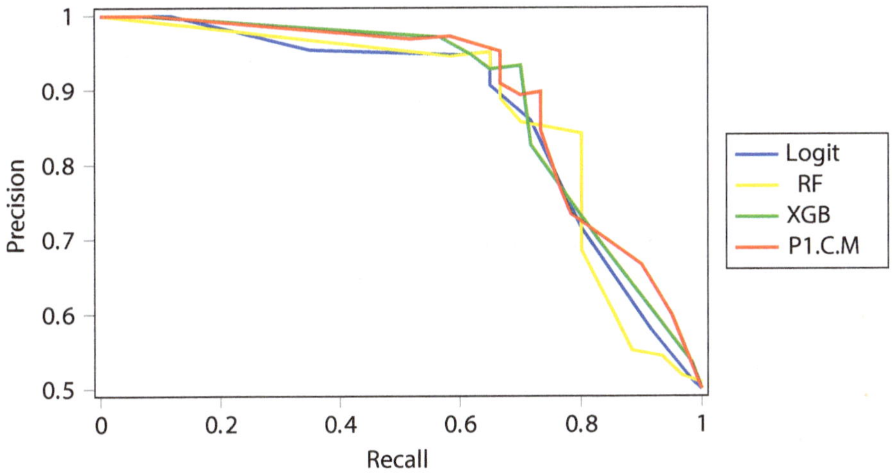

Fig. 4 Precision-Recall curves of the pipelines Logit, RF, XGB, and P1.C.M, using the ML_Dataset

performances are shown in Table 6, which also reports for each pipeline the area under the Precision-Recall (PR) curve. According to Pigini (2021), the PR curve is preferred over the ROC curve when dealing with imbalanced datasets (which is the majority of cases in EWSs). Even though the dataset used in this work is not imbalanced, the PR curve was chosen to facilitate comparisons with other works.

The P1.C.M and P3.C.R pipelines have the best F1 scores (within the context of the ML_Dataset and ML_Rolling_Dataset). It was expected as the result of the optimization described in Sect. 4. More details in Figs. 4 and 5.

Even though the analysis of the pipeline output shows that the P3.C.R and P1.C.M pipelines perform slightly better than the state-of-the-art considered as the reference baseline(s), the performance gain is minimal and falls within the range of variabil-

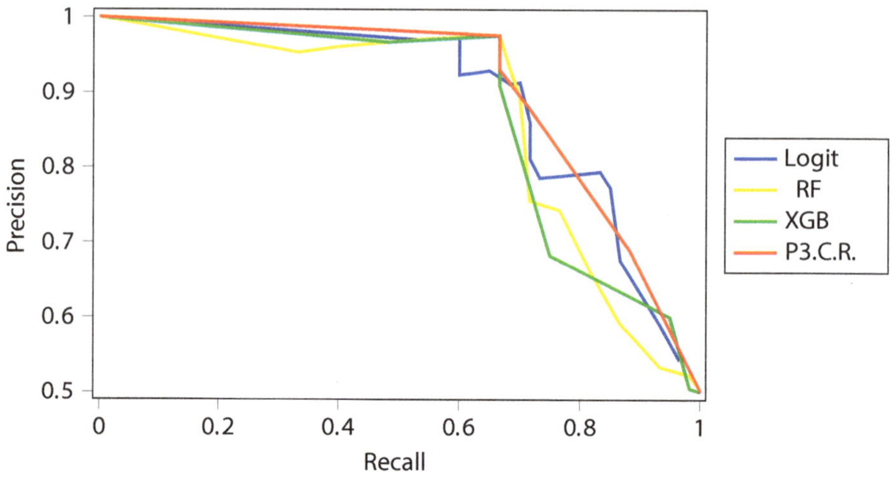

Fig. 5 Precision-Recall curves of the pipelines Logit, RF, XGB, and P3.C.R., using the ML_Rolling_Dataset

ity in the performance (i.e., training the same classifiers multiple times can still yield slightly different performance, a well-known issue in machine learning). Furthermore, the work performed to build the ML_Dataset and ML_Rolling_Dataset undoubtedly had a positive impact on both the proposed approach and the literature reference models. Since most of the literature focuses on imbalanced datasets, this work is not directly comparable to those approaches. Therefore, the performance contribution of Auto ML in the EWSs for financial crises has to be better evaluated in future work.

It is worth noting that the genetic programming optimization performed by the `TPOT` library may result in a local minimum; therefore, the pipelines identified in this chapter may be further improved. Generally speaking, improving the number of generations or working on a larger dataset can improve the stability of genetic algorithm searches. Unfortunately, in this scenario, it was not possible to further extend the train set (for example, at the expense of the test), because doing so would have made both datasets imbalanced or not anymore representative.

4.5 SHAP Values

The Shapley Additive exPlanations (SHAP) method was used to investigate the pipeline models P1.C.M and P3.C.R presented in Sects. 4.2 and 4.3. This model-agnostic method, based on cooperative game theory, provides consistent and locally accurate feature attributions for individual predictions (Lundberg & Lee, 2017). SHAP values quantify the contribution of each input feature to the model's output by

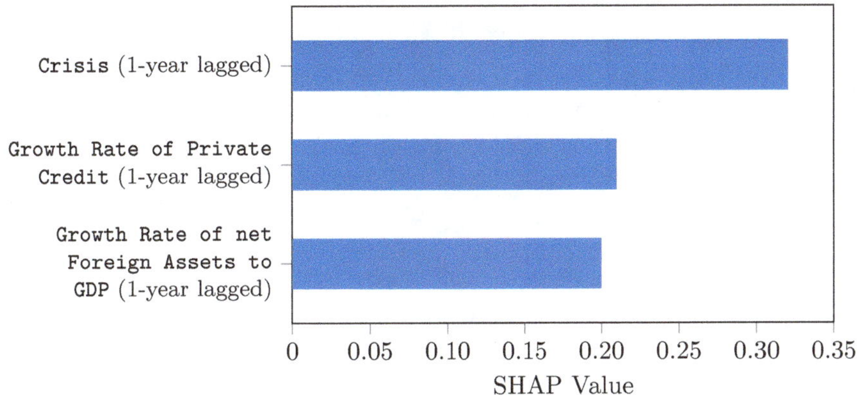

Fig. 6 Barchart showing the Mean Absolute SHAP values computed using the P1.C.M pipeline on the ML_Dataset (the test set was used)

computing the average marginal contribution of a feature across all possible feature subsets. This approach enables both global interpretability (by aggregating feature contributions across the dataset) and local interpretability (by explaining individual predictions), offering a comprehensive understanding of model behavior and feature importance.

Figures 6 and 7 show the mean absolute SHAP values computed using the P1.C.M and P3.C.R pipelines, respectively, on the ML_Dataset and ML_Rolling_Dataset. Specifically, the SHAP values shown were calculated on the test set.

Figure 6 shows the features that contribute most to crises prediction using the pipeline P1.C.M trained on the ML_Dataset; in particular they are

- One-year lagged `Crisis`. As expected, the occurrence of a crisis in the previous year provides useful information that increases the likelihood of predicting a future crisis.
- One-year lagged `Growth Rate of Private Credit`. It measures the growth of real domestic credit. It is expected to have a positive effect on the probability of a crisis, which is likely to increase with over-indebtedness and deterioration in the quality of bank assets.
- One-year lagged `Growth Rate of net Foreign Assets to GDP`. It tracks the evolution of the country's external position. A decline in this ratio is generally expected to increase the probability of a crisis, as it signals widening external imbalances, potential capital flight, or a reduced capacity to absorb external shocks.
- The other features have less evident effects.

Figure 7 shows the features that contribute most to crises prediction using the pipeline P3.C.R trained on the ML_Rolling_Dataset. It is worth recalling that, as outlined in Sect. 3.4, features in this dataset have been enriched with previous years'

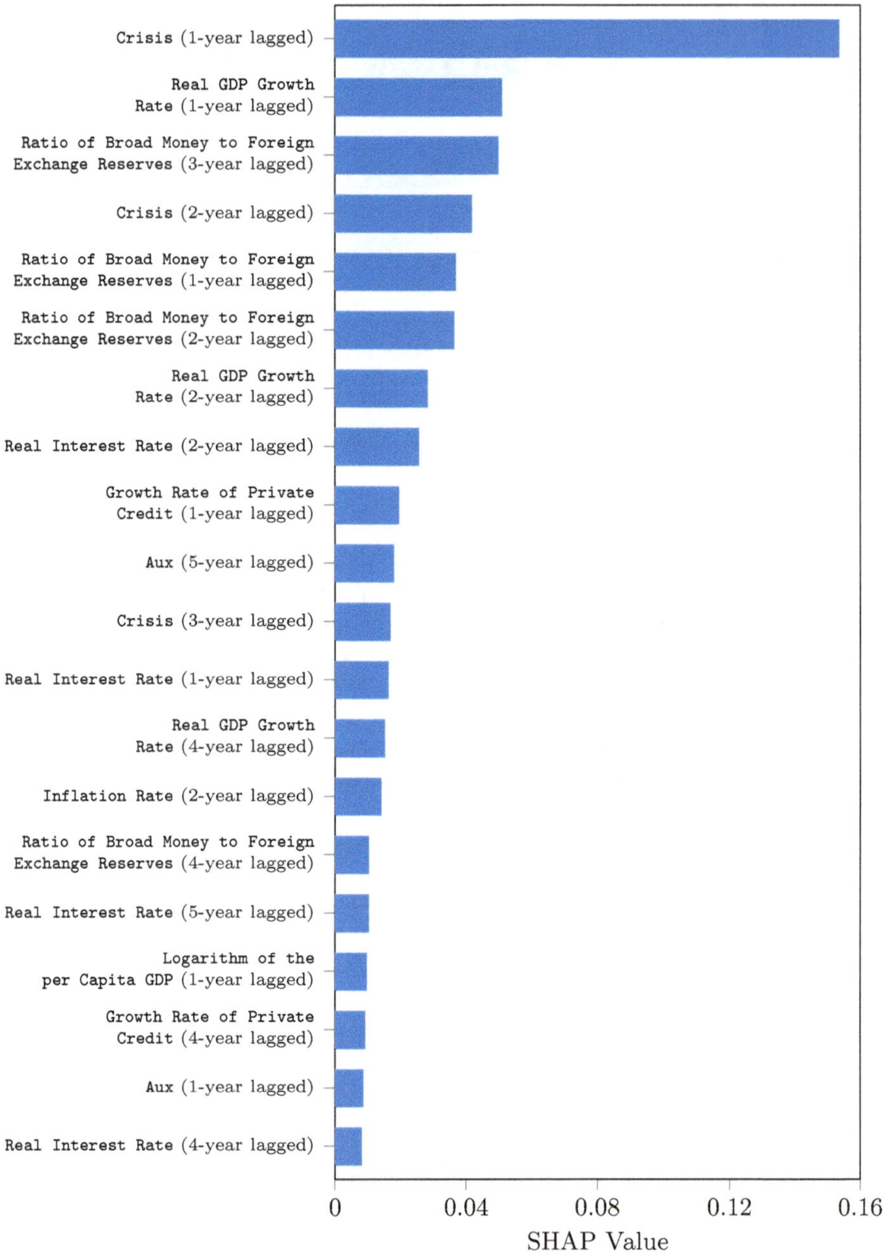

Fig. 7 Barchart showing the Mean Absolute SHAP values computed using the P3.C.R pipeline on the ML_Rolling_Dataset (the test set was used)

values. The most important features, according to the mean absolute SHAP values, are the following (only the first ones are commented on):

- One-year and two-year lagged `Crisis`. As already discussed, past crisis occurrences provide useful information to predict future instabilities.
- One-year lagged `Real GDP Growth Rate`. Lower values of this variable are typically associated with weak macroeconomic conditions and may indicate increasing financial stress.
- Three- and two-year lagged `Ratio of Broad Money to Foreign Exchange Reserves`. These are monetary variables, which, as expected, positively affect the probability of predicting a crisis.

As outlined in Fig. 7, the addition of previous years' data has been taken into account by the classifier. This can explain the (slightly) better performance of the P3.C.R. over the P1.C.M pipeline.

4.6 Considerations

Different runs of `TPOT` (on the same dataset and element pools) are likely to produce different pipelines for several reasons. For instance, multiple pipelines may achieve comparable top performance (although increasing the number of generations could further differentiate them); most pipeline elements are randomly initialized; and even when a seed is specified, training the same pipeline multiple times can still yield slightly different performance, a well-known issue in machine learning. Nevertheless, this does not represent a significant problem, as the pipelines identified by Auto ML tools are intended to serve as starting points rather than final outcomes of a research effort.

5 Conclusions and Future Work

This work has focused on ML methods to show their potential. Statistical models and ML methods are complementary, often overlapping and enriching each other. Statistical models, such as those presented in different chapters of this book, are based on assumptions and aim to estimate parameters that can be accurately interpreted to explain associations between variables, including inference, while also providing measures of uncertainty through confidence intervals, p-values, and hypothesis tests. Moreover, thanks to these assumptions, they remain stable with small datasets and are robust because their simplicity and constraints often allow them to remain valid with new data, as long as the assumptions hold.

As illustrated in this chapter, the ML approach is oriented toward building predictive models by learning from data without necessarily requiring specific assumptions. ML algorithms, such as those applied here, can capture complex, non-linear patterns without the user defining an explicit model structure. However, the statistical literature (Bishop, 2006; Giudici, 2003; Hastie et al., 2009; James et al., 2013) documents several limitations—especially for complex models such as neural networks or ensembles (e.g., random forests, XGBoost)—they are often "black boxes", difficult to interpret, and more focused on prediction than on inference. Non-parametric models, in particular, require large datasets to perform well, are prone to overfitting without proper regularization and validation, and can be less robust in new contexts or when applied to data from different distributions. In addition, some algorithms (e.g., deep learning and boosting) are computationally more demanding to train and validate than simpler models.

Two possible paths, therefore, emerge. One is to choose between the approaches: ML is particularly suitable for complex data structures because it does not require explicit assumptions, whereas when data are limited and accurate, and an explainable interpretation is needed, statistical models are often preferable for their interpretability and robustness. The other path is to combine the two in hybrid approaches. Statistical models provide a solid basis for analysis and interpretation, while ML can refine predictions. For example, statistical models can assess variable importance within an ML framework, or Bayesian ML methods—such as Bayesian optimization and probabilistic graphical models—can integrate both paradigms to improve predictions, especially under uncertainty or incomplete data. Statistical techniques can also prepare data for ML (e.g., normalization, principal component analysis, variable selection), while statistical methods such as Bayesian inference or cross-validation can optimize ML hyperparameters. Applications include health econometrics, where the goal is to combine the predictive power of ML methods with interpretability (Athey & Imbens, 2017; Chernozhukov et al., 2018).

Appendix A

The algorithm used for undersampling and train-test split is shown in Listing 1.

Appendix B

In this appendix the pipelines not previously shown are reported next.

B.1 Pipeline P1.C.M

The P1.C.M, i.e., the best pipeline that `TPOT` identified working on the Custom_Pool, of elements and on ML_Dataset, is reported in Listing 5.

Algorithm 1 Undersampling and train-test split using *sel* as stratification variable

```
Print "Performing undersampling and train & test split"
Print "The initial dataset is stored in the df dataframe"
Print "A new variable df_group will be added, accepting the values 'balanced_train', 'balanced_test', 'test2', and 'test3'. The initial default value is 'test3'. "
Let testSizePerc = 0.3 # Test size is 30%
for all group g and frequency f in value_counts of df.sel do
    Let df_group ← subset of df where sel == g
    Let c1 ← number of rows in df_group where crisis0 == 1
    Let c0 ← number of rows in df_group where crisis0 == 0
    Let m ← min(c0, c1) # m is the size of the minority class
    if m == 0 then # Either the crisis or non-crisis cases are missing
        Assign all rows in df_group to group 'test3' # These records would be completely dropped by undersampling. They are labelled test3 because they can later be used as an additional test set
    else if m == 1 then # If the minority class has size 1, it cannot be a train-test split, so it is assigned to the train subset
        for target in {0, 1} do # in both crisis and non-crisis subsets
            Randomly sample 1 row from df_group where crisis0 == target
            Assign sampled row to group 'balanced_train'
        end for
    else # The minority class has at least 2 records (at least one for the train set and one for the test set)
        Assign all rows in df_group to group 'test2' # test2 is the initial value at the beginning
        Let t ← round(m · testSizePerc) # t is the test set size of the minority class
        Let t ← max(1, t) # if t is lower than 1, then it is set to 1.
        Let r ← m − t # train set size on the minority class
        for target in {0, 1} do
            Randomly sample m rows from df_group where crisis0 == target
            Randomly sample t rows from the m rows
            Assign the sampled t rows to group 'balanced_test'
            Assign the remaining m rows to group 'balanced_train' # All labels assigned neither to balanced_test' nor to balanced_train' will keep the initial 'test2' label. At the end of the process, these are the records dropped by undersampling. They are labeled test2 because they can later be used as an additional test set.
        end for
    end if
end for
```

The `make_pipeline` class creates a pipeline instance. Its parameters are the pipeline elements, in the exact order they will occupy the pipeline. In Listing 5, the pipeline is composed of two elements. The first is the `MakeUnion` component, which creates parallel branches in a pipeline. The same input is passed to different parallel branches, which are defined when the `MakeUnion` class is instantiated. In this case, the `MakeUnion` function has two branches: the first, named `StackingEstimator`, is a class provided by `TPOT` for creating pipelines involving multiple classifiers. It is a meta-transformer that wraps a classifier and adds its predictions and/or class probabilities as synthetic feature(s) to the input dataset. When the `StackingEstimator` is created, a classifier is specified and stored

```
from sklearn.pipeline import make_pipeline, make_union
from sklearn.tree import DecisionTreeClassifier
from sklearn.naive_bayes import BernoulliNB
from sklearn.preprocessing import FunctionTransformer
from copy import copy # it returns a shallow copy of the object
from sklearn.ensemble import ExtraTreesClassifier
from tpot.builtins import StackingEstimator

pipe=make_pipeline(
 make_union( # 1st pipeline element
  StackingEstimator(
  estimator=make_pipeline(
   StackingEstimator(
     estimator=BernoulliNB(alpha=10.0, fit_prior=True)),
    DecisionTreeClassifier(criterion="gini", max_depth=6, min_samples_leaf=11, min_samples_split=11))),
   FunctionTransformer(copy) # it copies the input features),
 ExtraTreesClassifier( # 2nd and last pipeline element
bootstrap=False, criterion="gini", max_features=0.05, min_samples_leaf=2, min_samples_split=13, n_estimators=100 ))
```

Listing 5 The P1.C.M Pipeline Code

internally. The `StackingEstimator` implements the `fit()`, `transform()`, and `predict()` methods required for pipelines, as outlined in Sect. 2.1. During prediction, the internal classifier processes the input features; its predictions are merged with the original input, and the combined result is passed to the next pipeline element. In our case, the `StackingEstimator` combines the predictions of two classifiers, a `BernoulliNB` and a `DecisionTreeClassifier`.

The second `MakeUnion` branch is the `FunctionTransformer(copy)` class, which creates a copy of the input features. The `MakeUnion` then merges the `DecisionTreeClassifier` results and the input features. Finally, the second pipeline element is the `ExtraTreesClassifier`, which uses the merge to produce the final prediction. The entire pipeline is trained on the training set and evaluated on the test set. The classification report obtained is shown in Table 7.

The use of `StackingEstimator` and `FunctionTransformer(copy)` inside the `make_union` produces a `make_union` output where the original input features are repeated twice. In fact, the `StackingEstimator` output contains the input features, and the `FunctionTransformer(copy)` copies the input

Table 7 Classification report of the pipeline P1.C.M. evaluated on the test set

	Precision	Recall	F1 score	Support
Non-crises (0)	0.76	0.83	0.79	63
Crises (1)	0.81	0.75	0.78	63
Accuracy	–	–	0.79	126
Average	0.79	0.79	0.79	126

Table 8 Classification report of the pipeline P2.L.M evaluated on the test set

	Precision	Recall	F1 score	Support
Non-crises (0)	0.74	0.90	0.81	63
Crises (1)	0.88	0.68	0.77	63
Accuracy	–	–	0.79	126
Average	0.81	0.79	0.79	126

features. Therefore, it is wise to manually restructure the pipeline as outlined in Listing 3.

The performance between P1.C.M and P1.C.M.Bis does not change much. However, tuning the last element (`ExtraTreesClassifier`) of the P1.C.M.Bis pipeline while keeping the preceding components unchanged could potentially improve classification performance. This evaluation will be considered in future studies.

B.2 Pipeline P2.L.M

The P2.L.M, i.e., the best pipeline that `TPOT` identified working on the Light_Pool of elements and on ML_Dataset is reported in Listing 6.
Most of the elements used in Listing 6 have already been introduced in Appendix B.1.
The comments should clarify how the new pipeline elements work.
The classification report is shown in Table 8.

B.3 Pipeline P4.L.R

The P4.L.R, i.e., the best pipeline that `TPOT` identified working on the Light_Pool of elements and on ML_Rolling_Dataset is reported in Listing 7.
Most of the elements used in Listing 7 have already been introduced in Appendix B.1.
The comments should clarify the new pipeline elements. The classification report is shown in Table 9.

```
from sklearn.pipeline import make_pipeline,
    make_union
from sklearn.preprocessing import
    FunctionTransformer
from sklearn.feature_selection import
    SelectPercentile
from sklearn.feature_selection import f_classif #
    # Compute the ANOVA F-value for the provided
    sample
from sklearn.preprocessing import Binarizer
from sklearn.naive_bayes import GaussianNB
from sklearn.neighbors import KNeighborsClassifier
from copy import copy

pipe=make_pipeline(
make_union( # 1st pipeline element
FunctionTransformer(copy),
make_pipeline(
StackingEstimator(estimator=GaussianNB()),
SelectPercentile( # Select features according to a
     percentile of the highest scores
score_func=f_classif, # see import
percentile=45
)
)
),
# 2nd pipeline element
Binarizer(threshold=0.25), # Binarize data (set
    feature values to 0 or 1) according to a
    threshold.
# 3rd and last pipeline element
KNeighborsClassifier(n_neighbors=13, p=2, weights=
    "distance")
)
```

Listing 6 The P2.L.M Pipeline Code

Table 9 Classification report of the pipeline P4.L.R evaluated on the test set

	Precision	Recall	F1 score	Support
Non-crises (0)	0.81	0.92	0.86	63
Crises (1)	0.91	0.78	0.84	63
Accuracy	–	–	0.85	126
Average	0.86	0.85	0.85	126

Appendix C

The (literature) baseline models described in Sect. 4.4 and the optimal hyperparameters found are outlined in Listings 8, 9, and 10.

```
from sklearn.pipeline import make_pipeline
from sklearn.preprocessing import Binarizer
from sklearn.tree import DecisionTreeClassifier

pipe=make_pipeline(
    Binarizer(threshold=0.45),
    DecisionTreeClassifier(criterion="gini",
        max_depth=4, min_samples_leaf=5,
        min_samples_split=4)
)
```

Listing 7 The P4.L.R Pipeline Code

```
logitPipeline = Pipeline([
('logreg', LogisticRegression(
max_iter=5000,
C=0.1,
penalty='l1',
solver='liblinear'
))
])
```

Listing 8 A baseline pipeline featuring a Logistic Regression classifier. The hyperparameters were selected via a grid search across both ML_Dataset and ML_Rolling_Dataset. The two different searches produced the same hyperparameters

```
rfPipe = Pipeline([
('clsf', RandomForestClassifier(
bootstrap=True,
max_depth=None, #=10 for the rolling dataset
min_samples_leaf=1,
min_samples_split=2,
n_estimators=100
))
])
```

Listing 9 A baseline pipeline featuring a Random Forest classifier. The hyperparameters were selected via a grid search across both ML_Dataset and ML_Rolling_Dataset. The two different searches produced similar but not equal hyperparameters. The differences are outlined using comments

```
xgbPipe = xgb.XGBClassifier(
objective='binary:logistic',
use_label_encoder=False,
eval_metric='logloss',
colsample_bytree=0.8,
gamma=1, # =0 for the rolling dataset
learning_rate=0.01,
max_depth=3,
min_child_weight=1, # =5 for the rolling dataset
n_estimators=100,
subsample=1.0   # = 0.8 for the rolling dataset
)
```

Listing 10 An XGBoost classifier. The hyperparameters were selected via a grid search across both ML_Dataset and ML_Rolling_Dataset. The two different searches produced similar but not equal hyperparameters. The differences are outlined using comments

References

Athey, S., & Imbens, G. (2017). The state of applied econometrics: Causality and policy evaluation. *Journal of Economic Perspectives, 31*, 3–32.

Azedou, A., Amine, A., Kisekka, I., & Lahssini, S. (2025). Genetic algorithm optimization of ensemble learning approach for improved land cover and land use mapping: Application to talassemtane national park. *Ecological Indicators, 177*, 1–14.

Azure. (2025). Automated machine learning. https://azure.microsoft.com/en-us/solutions/automated-machine-learning. Retrieved August 20, 2025

Bishop, C. M. (2006). *Pattern Recognition and Machine Learning*. New York: Springer.

Breiman, L. (2001). Random forests. *Machine learning, 45*, 5–32.

Casabianca, E. J., Catalano, M., Forni, L., Giarda, E., & Passeri, S. (2022). A machine learning approach to rank the determinants of banking crises over time and across countries. *Journal of International Money and Finance, 129*, 1–24.

Chawla, N. V., Bowyer, K. W., Hall, L. O., & Kegelmeyer, W. P. (2002). SMOTE: Synthetic minority over-sampling technique. *Journal of Artificial Intelligence Research, 16*, 321–357.

Chen, T., & Guestrin, C. (2016). XGboost: A scalable tree boosting system. In C. Aggarwal, A. Smola, R. Rastogi, D. Shen, B. Krishnapuram, & M. Shah (Eds.), *Proceedings of the 22nd ACM SIGKDD International Conference on Knowledge Discovery and Data Mining, KDD '16* (pp. 785–794). Association for Computing Machinery.

Chernozhukov, V., Chetverikov, D., Demirer, M., Duflo, E., Hansen, C., Newey, W., & Robins, J. (2018). Double/debiased machine learning for treatment and structural parameters. *The Econometrics Journal, 21*, C1–C68.

Domingos, P. (2012). A few useful things to know about machine learning. *Communications of the ACM, 55*, 78–87.

EpistasisLab (2025). Tpot documentation. https://github.com/EpistasisLab/tpot. Retrieved August 20, 2025.

Feurer, M., Klein, A., Eggensperger, K., Springenberg, J., Blum, M., & Hutter, F. (2015). Efficient and robust automated machine learning. *Advances in Neural Information Processing Systems, 28*, 2962–2970.

Giudici, P. (2003). *Applied data mining: Statistical methods for business and industry*. Wiley.

Google Cloud (2025). *Cloud AutoML*. https://cloud.google.com/automl. Accessed 20 Aug, 2025..

Hastie, T., Tibshirani, R., & Friedman, J. (2009). *The elements of statistical learning. Data mining, inference, and prediction* (2nd ed.). Springer.

He, H., & Garcia, E. A. (2009). Learning from imbalanced data. *IEEE Transactions on Knowledge and Data Engineering, 21*, 1263–1284.

He, X., Zhao, K., & Chu, X. (2021). AutoML: A survey of the state-of-the-art. *Knowledge-Based Systems, 212*, 1–27.

James, G., Witten, D., Hastie, T., & Tibshirani, R. (2013). *An introduction to statistical learning*. New York: Springer.

Jin, H., Song, Q., & Hu, X. (2019). Auto-keras: An efficient neural architecture search system. *Proceedings of the 25th ACM SIGKDD International Conference on Knowledge Discovery and Data Mining* (pp. 1946–1956).

Krawczyk, B. (2016). Learning from imbalanced data: Open challenges and future directions. *Progress in Artificial Intelligence, 5*, 221–232.

LeCun, Y., Bengio, Y., & Hinton, G. (2015). Deep learning. *Nature, 521*, 436–444.

Leung, F.H.-F., Lam, H.-K., Ling, S.-H., & Tam, P.K.-S. (2003). Tuning of the structure and parameters of a neural network using an improved genetic algorithm. *IEEE Transactions on Neural Networks, 14*, 79–88.

Li, L. (2025). Research on the construction of international financial crisis early warning model based on xgboost algorithm. In M. Zhang, N. A. Y. Hui, A. T. Rahman, & K. Shehzad (Eds.), *Proceedings of the 2nd Guangdong-Hong Kong-Macao Greater Bay Area International Conference on Digital Economy and Artificial Intelligence, DEAI '25* (pp. 1335–1339). Association for Computing Machinery.

Lipton, Z. C. (2018). The mythos of model interpretability: In machine learning, the concept of interpretability is both important and slippery. *Queue, 16*, 31–57.

Lundberg, S. M., & Lee, S.-I. (2017). A unified approach to interpreting model predictions. *Advances in Neural Information Processing Systems, 30*, 4765–4774.

Mahalle, A. V., et al. (2024). Machine learning algorithms for data-driven intelligent systems. In M. Ghonge, et al. (Eds.), *Data-driven systems and intelligent applications* (pp. 52–61). Taylor & Francis.

McCullagh, P., & Nelder, J. A. (2019). *Generalized linear models*. Boca Raton, FL: Routledge.

Mitchell, M. (1998). *An introduction to genetic algorithms*. Cambridge, MA: MIT Press.

Nazareth, N., & Reddy, Y. V. R. (2023). Financial applications of machine learning: A literature review. *Expert Systems with Applications, 219*, Article 119640.

Nielsen, M. A. (2015). *Neural networks and deep learning*. San Francisco, CA: Determination Press.

Olson, R. S., & Moore, J. H. (2016, 24 Jun). TPOT: A tree-based pipeline optimization tool for automating machine learning. In F. Hutter, L. Kotthoff, & J. Vanschoren (Eds.), *Proceedings of the Workshop on Automatic Machine Learning, Volume 64 of Proceedings of Machine Learning Research* (pp. 66–74). PMLR.

Olson, R. S., et al. (2016). Evaluation of a tree-based pipeline optimization tool for automating data science. *In Proceedings of the Genetic and Evolutionary Computation Conference, 2016*, 485–492.

Pedregosa, F., et al. (2011). Scikit-learn: Machine learning in python. *The Journal of Machine Learning Research, 12*, 2825–2830.

Perez, E., Kiela, D., & Cho, K. (2021). True few-shot learning with language models. *Advances in Neural Information Processing Systems, 34*, 11054–11070.

Pigini, C. (2021). Penalized maximum likelihood estimation of logit-based early warning systems. *International Journal of Forecasting, 37*, 1156–1172.

Python Software Foundation. (2025). The python language reference. https://docs.python.org/3/reference/index.html. Retrieved August 20, 2025.

Qcut. (2025). Quantile based discretization function. https://pandas.pydata.org/docs/reference/api/pandas.qcut.html. Retrieved August 20, 2025.

Real, E., Aggarwal, A., Huang, Y., & Le, Q. V. (2019). Regularized evolution for image classifier architecture search. In *Proceedings of the AAAI conference on artificial intelligence* (Vol. 33, pp. 4780–4789).

Schmidhuber, J. (2015). Deep learning in neural networks: An overview. *Neural Networks, 61*, 85–117.

Snoek, J., Larochelle, H., & Adams, R. P. (2012). Practical bayesian optimization of machine learning algorithms. *Advances in Neural Information Processing Systems, 25*, 1–9.

van Rossum, G., & Drake, F. L. (2011). *The python language reference manual*. Network Theory Ltd.

Yaloveha, V., Podorozhniak, A., & Kuchuk, H. (2022). Convolutional neural network hyperparameter optimization applied to land cover classification. *Radioelectronic and computer systems, 101*, 115–128.

Zoph, B., & Le, Q. V. (2017). Neural architecture search with reinforcement learning. In *International Conference on Learning Representations* (pp. 1–16).

Exploring Binary Regression and Hidden Markov Models for Early Warning Systems

Luca Brusa, Fulvia Pennoni, Romina Peruilh Bagolini, and Francesco Bartolucci

Abstract Short-term forecasting has gained increasing attention across various fields, driven by the growing need for model-based probabilistic predictions to support early warning systems. We introduce several statistical approaches for probabilistic forecasting of a binary response, namely, multiple probit and logit regression, logit fixed-effects models, and hidden Markov models. We compare in-sample and out-of-sample predictions based on longitudinal data on macroeconomic and financial indicators for 129 countries over the period 1983–2017, using a binary indicator for the occurrence of a financial crisis in a given year. The lagged response is included as a predictor of crises, together with other covariates, and missing data are handled within the hidden Markov model framework. The results are evaluated using a range of metrics and demonstrate the effectiveness of the proposed modeling approaches.

Keywords Forecasting · Missing values · Panel data · Stochastic models · Rare events

---

Supplementary Information The online version contains supplementary material available at https://doi.org/10.1007/978-3-032-24141-2_3.

L. Brusa (✉) · F. Pennoni
University of Milano-Bicocca, Milan, Italy
e-mail: luca.brusa@unimib.it

F. Pennoni
e-mail: fulvia.pennoni@unimib.it

R. Peruilh Bagolini · F. Bartolucci
University of Perugia, Perugia, Italy
e-mail: romina.peruilh@studenti.unipg.it

F. Bartolucci
e-mail: francesco.bartolucci@unipg.it

F. Bartolucci et al. (eds.), *Models for Longitudinal Data with Applications to Early Warning Systems*, Springer Proceedings in Mathematics & Statistics 553,
https://doi.org/10.1007/978-3-032-24141-2_3

1 Introduction

In the past few years, financial global stability has been subject to shocks from adverse events such as the COVID-19 pandemic, international tensions, and energy market disruptions. These events have reinforced the need for analytical frameworks to detect early signals of distress, allowing the mitigation of potential hazards. These frameworks, usually referred to as early warning systems (EWSs), typically rely on statistical models capable of analyzing data and detecting early signs of abnormal behavior, providing actionable insights for timely interventions. The conceptual foundations of model-based EWS can be traced back to the late 1970s and early 1990s, motivated by episodes of currency crises in Latin America and the financial crises in Mexico and Asia.

Nowadays, the use of statistical models as EWSs has become an integral part of surveillance in many fields as they play a crucial role in recognizing patterns, anomalies, and signals that may indicate the onset of adverse events. Health-related EWSs have been instrumental in tracking the spread of infectious diseases such as COVID-19 (see Dasgupta & Wheeler, 2020), facilitating the timely implementation of public health measures. Climate-related EWSs are increasingly adopted in the attempt to predict natural disasters, such as tropical cyclones, floods, droughts, heatwaves, landslides, volcanoes, earthquakes, and tsunamis, improving disaster preparedness and quality of responses (Basher, 2006; Umar et al., 2014). Other common applications of EWS are connected with software development (Malhotra et al., 2022), customer retention (Rothenbuehler et al., 2015), and autonomous driving technology (David & Söffker, 2023). As the importance of EWSs continues to grow, the attempts to further improve their effectiveness also increase. A key challenge is ensuring the ongoing refinement of EWS models to evolving real-world scenarios. Factors such as data quality, key predictors, and model complexity may impact the predictive performance of such systems. Consequently, continuous research to refine these models is necessary to enhance their effectiveness.

In this chapter, we explore the use of various predictive models and their application as EWS in a real-world scenario. The exposition begins with an introduction to binary regression models and progresses toward more advanced approaches tailored to the analysis of longitudinal data, such as the hidden Markov (HM) models (Bartolucci et al., 2022; Mews et al., 2025) based on discrete latent variables. We highlight their use in the context of in-sample prediction and out-of-sample forecasting, illustrating the computation of predicted response probabilities and the choice of suitable thresholds to classify them into the predicted response categories. We show how this procedure can be employed as an EWS. In addition, we discuss the interpretability of the estimated results from the models under consideration and compare their predictive performance using a range of evaluation metrics. Following the proposal in Pigini (2021) and using the same data on banking crises, we aim to assess the characteristics and peculiarities of supervised modeling approaches for early warnings, extending the previous work to the advanced family of HM models and providing useful comparison among the proposed approaches. In this dataset, a systemic banking crisis is recorded when, for a given country and year, there are indications of financial distress in the banking system (such as bank runs, losses, or liquidations) and policy interventions have been implemented in response to significant losses (Laeven & Valencia, 2018). Three categories of variables, commonly used in the literature as predictors (see Caggiano et al., 2016; Davis & Karim, 2008; Demirgüç-Kunt & Detragiache, 2005), are also available: the first is related to macroeconomic information, the second includes monetary variables, and the third comprises financial information. The code employed for developing the application has been implemented for the software R (R Core Team, 2025).

The remainder of this chapter is organized as follows. Section 2 provides a brief overview of the literature regarding regression and HM models used as EWS and describes the main performance metrics for these models. Section 3 discusses the use of binary response models as EWS, detailing their model formulations. Section 4 illustrates the proposed formulation and estimation of the HM model, along with the in-sample predictive distribution and the out-of-sample forecasting distribution. Section 5 presents the results of proposed model-based EWS for analyzing financial crises, first reporting those of the logit and probit models, followed by those of the HM model. Section 6 provides some concluding remarks. Appendix A contains additional details related to the HM model results.

2 Preliminaries

In the following, we summarize some basic features of model-based EWSs. Section 2.1 outlines key contributions from the literature related to binary regression and HM models, while Sect. 2.2 reviews the main metrics commonly used to evaluate the predictive performance of these approaches.

2.1 Model-Based Approaches for Early Warning Systems

Binary regression models have been recurrently used as data-driven EWSs. Pooled logit-based EWS have shown good predictive performance for banking crises compared to non-parametric approaches (Davis & Karim, 2008). Extended versions of pooled probit-based EWS have allowed to consider the crisis duration when crises persist over longer periods (Antunes et al., 2018). Fixed-effects model specifications have enabled to extract additional information when dealing with longitudinal data, thereby enhancing predictive power. In particular, the fixed-effects logit model, estimated by a penalized maximum likelihood approach, has been shown to outperform pooled models, especially when the event of interest occurs in only a small proportion of sample units.

HM models (Bartolucci et al., 2013) are another family of models used as EWS because, by modeling the underlying dynamics of latent processes, they provide additional insights. In particular, they play an important role in capturing time-varying unobserved heterogeneity, which is crucial in the analysis of repeated measurements. Among others, Jiang et al. (2016) employed a continuous HM model to provide early warning signals of risk for algal blooms in water systems, while Yip et al. (2018) proposed an EWS to forecast the magnitude and timing of earthquakes through an HM model. Further, Chen et al. (2016) implemented an HM model-based approach to generate early warning signals to a transitional phase before a critical state during complex diseases as acute lung injury, breast cancer, HCV-induced dysplasia, and hepatocellular carcinoma. Bharati & Lomte (2016) used HM models as prediction systems to raise early warnings about potential cloud attacks in which the hidden variable represents the overall security state of the cloud. Other proposals of HM models for EWSs are related to the prediction of international political crises using demographic and military data (Schrodt, 1997), as well as the identification of business cycle states using macroeconomic indicators (Gregoir & Lenglart, 2000). Overall, nowadays, model-based EWSs, such as binary regression and HM models, play an important role in reducing uncertainty and increasing preparedness to adverse events across a wide range of areas.

2.2 *Predictive Accuracy and Beyond: Evaluation Metrics*

Different metrics can be used to assess the predictive performance of machine learning and statistical models. This performance can be evaluated either by analyzing the predicted responses or by directly examining the predicted probabilities. In the first case, the simplest way to assess model performance is by using a confusion matrix, which provides a cross-tabulation of observed and predicted classes, identifying true positives (TP), false positives (FP), false negatives (FN), and true negatives (TN), according to the assumed model. Based on these values, additional performance metrics can be derived. Table 1 summarizes the main metrics commonly used for a binary response. Accuracy provides a global assessment of predictive performance by measuring the proportion of correctly classified instances. However, the accuracy metric can be misleading when datasets exhibit an imbalanced response distribution. In such case, indeed, high accuracy values can be achieved by favoring the majority class while failing to correctly classify the minority class. This issue, known as the accuracy paradox (Valverde-Albacete & Peláez-Moreno, 2014), highlights the need for using additional evaluation metrics. Sensitivity (or recall) quantifies the ability of the model to correctly identify positive instances, whereas specificity measures its ability to correctly recognize true negative cases. Precision represents the proportion of correctly identified positive cases among the instances predicted as positive. It is important to highlight the trade-offs among these metrics: improvements in sensitivity often come at the expense of specificity or precision. Determining which metric to emphasize should be guided by the particular goals of the analysis and the relative costs of false positives and false negatives.

Table 1 Formulas and definitions of the most commonly used evaluation metrics of predictive model performance

	Formula	Definition
Basic metrics		
Accuracy	$\frac{TP + TN}{TP + FP + TN + FN}$	Proportion of correctly classified instances, which reflects the overall predictive ability of the model
Sensitivity (Recall)	$\frac{TP}{TP + FN}$	Proportion of actual positives correctly predicted, which is relevant when the focus is on identifying positive cases
Specificity	$\frac{TN}{TN + FP}$	Proportion of actual negatives correctly predicted, which is important to control false alarms
Precision	$\frac{TP}{TP + FP}$	Proportion of correct positive predictions, which is relevant when false positives are costly
Derivative metrics		
Youden's index	Sensitivity + Specificity − 1	Overall model effectiveness, which is robust for class-imbalanced datasets
F1-score	$\frac{\text{Precision} \times \text{Recall}}{0.5(\text{Precision} + \text{Recall})}$	Weighted harmonic mean of precision and recall, which is preferable for class-imbalanced datasets

When dealing with predicted response probabilities, Receiver Operating Characteristic (ROC, Fawcett, 2006) and Precision–Recall (PR, Fawcett, 2006) analyses are widely employed. These metrics assess performance across a continuum of thresholds $c \in [0, 1)$, where each value of c defines a (potentially) different classification of predicted probabilities into positive and negative outcomes. The ROC curve plots sensitivity against one minus specificity across these thresholds, providing a graphic representation of the discriminatory capacity of a model. The area under the ROC curve (AUROC) serves as an overall performance measure, ranging from 0 to 1, where 0.5 indicates random classification and 1 represents perfect classification (Hanley & McNeil, 1982). Youden's index (Youden, 1950), denoted as J and defined as sensitivity + specificity − 1, is a widely used measure derived from the ROC curve. Its values range from −1 to 1, with higher values indicating greater discriminatory ability. Graphically, this index represents the vertical distance between the curve and the diagonal bisector, and may be useful to determine the optimal threshold for converting predicted probabilities into predicted responses. The optimal threshold, denoted by $\tilde{c}$, is obtained as $\tilde{c} = \text{argmax}_c\, J(c)$, optimizing the trade-off between sensitivity and specificity. ROC analysis is a popular approach for assessing binary classification models; however, special attention is required when it is applied to imbalanced datasets (Berrar & Flach, 2012).

An alternative or complementary measure suitable to assess model performance in imbalanced settings is the PR curve (Davis & Goadrich, 2006; Saito & Rehmsmeier, 2015). Unlike ROC analysis, which evaluates the overall discriminatory ability, the PR curve focuses on the trade-off between correctly identifying positive instances, minimizing false positives. The area under the PR curve (AUPR) ranges from 0 to 1, where, in general, values above 0.8 indicate good performance results. The F1-score (Davis & Goadrich, 2006), closely related to the PR curve, is defined as the harmonic mean of precision and recall. Similar to Youden's index, the F1-metric can be optimized across different threshold values to identify the optimal one, denoted by $\tilde{c} = \text{argmax}_c\, F1(c)$, and may be chosen to transform predicted probabilities into predicted responses.

3 Logit and Probit Models

Regression models for binary responses represent a class of simple, yet effective, statistical methods widely used for forecasting. In the context of EWSs, these models provide both interpretability and accuracy. In cross-sectional frameworks, common response model specifications include the logit and probit models (Agresti, 2013), which differ in the assumed distribution of the error terms. In longitudinal contexts, additional specifications, such as pooled or fixed-effects (Wooldridge, 2010), can be adopted. Pooled models assume common associations across the individuals in the sample, whereas fixed-effects models account for individual-specific characteristics that remain constant over time; see, among others, Heckman (1981), Antunes et al. (2018), and Hsiao (2022).

3.1 Main Assumptions

Referring to longitudinal data, let $Y_i^{(t)}$ denote the binary response for unit i at time occasion t, $i = 1, \ldots, n$, $t = 1, \ldots, T$. In this way, $Y_i^{(t)} = 1$ represents the occurrence of the event of interest for unit i at time t. Moreover, in the presence of individual covariates, we let $\boldsymbol{X}_i = (\boldsymbol{X}_i^{(1)}, \ldots, \boldsymbol{X}_i^{(T)})$, with $\boldsymbol{X}_i^{(t)}$ representing the vector of covariates for unit i at time t. In the following, $\boldsymbol{y}_i$ and $\boldsymbol{x}_i = (\boldsymbol{x}_i^{(1)}, \ldots, \boldsymbol{x}_i^{(T)})$ denote realizations of $\boldsymbol{Y}_i$ and $\boldsymbol{X}_i$, respectively.

Under the class of dynamic binary regression models (Agresti, 2013; McCullagh & Nelder, 1989), the conditional response probability, denoted as

$$p_i^{(t)} = p(Y_i^{(t)} = 1 | \boldsymbol{X}_i^{(t)} = \boldsymbol{x}_i^{(t)}, Y_i^{(t-1)} = y_i^{(t-1)}),$$

is modeled as a function of the explanatory variables, which depends on the specific model, and it is known as response function.

The dynamic logit model assumes the response function to be the cumulative distribution function (CDF) of the standard logistic distribution $\Lambda(\cdot)$:

$$p_i^{(t)} = \Lambda(\beta_0 + \boldsymbol{\beta}'\boldsymbol{x}_i^{(t)} + \gamma\, y_i^{(t-1)}) = \frac{\exp(\beta_0 + \boldsymbol{\beta}'\boldsymbol{x}_i^{(t)} + \gamma\, y_i^{(t-1)})}{1 + \exp(\beta_0 + \boldsymbol{\beta}'\boldsymbol{x}_i^{(t)} + \gamma\, y_i^{(t-1)})}, \tag{1}$$

where β_0 is the intercept, $\boldsymbol{\beta}$ is the vector of regression coefficients, and γ is the slope parameter associated with the lagged response. The parameter γ captures the crisis state dependence, that is, the effect for country i at time t on the probability of experiencing a crisis at time t, given to the crisis status of the country at time $t-1$. The response function specifies the connection between the probability of $Y_i^{(t)} = 1$ and the linearly related explanatory variables along with the lagged response, and it allows us to restrict the response probabilities between 0 and 1 in Eq. (1).

Alternatively, binary responses can be modeled by the probit model. Assuming that the response function corresponds to the standard normal CDF, $\Phi(\cdot)$, the probit model is as follows:

$$p_i^{(t)} = \Phi(\beta_0 + \boldsymbol{\beta}'\boldsymbol{x}_i^{(t)} + \gamma\, y_i^{(t-1)}) = \int_{-\infty}^{\beta_0 + \boldsymbol{\beta}'\boldsymbol{x}_i^{(t)} + \gamma\, y_i^{(t-1)}} \phi(s)ds, \tag{2}$$

where $\phi(\cdot)$ denotes the probability density function of the standard Gaussian distribution, and β_0, $\boldsymbol{\beta}$, and γ are defined as in Eq. (1).

Both logit and probit models assume identical intercept β_0 and slope coefficients $\boldsymbol{\beta}$ and γ for all the observed units and periods. As no assumption on individual heterogeneity is made, they are also referred to as pooled models. A suitable approach to consider the individual-specific differences is the dynamic fixed-effects logit model. In this case, the conditional probability of $Y_i^{(t)} = 1$ given $\boldsymbol{X}_i^{(t)} = \boldsymbol{x}_i^{(t)}$ and $Y_i^{(t-1)} = y_i^{(t-1)}$ is as follows:

$$p_i^{(t)} = \Lambda(\alpha_i + \boldsymbol{\beta}'\boldsymbol{x}_i^{(t)} + \gamma\, y_i^{(t-1)}) = \frac{\exp(\alpha_i + \boldsymbol{\beta}'\boldsymbol{x}_i^{(t)} + \gamma\, y_i^{(t-1)})}{1 + \exp(\alpha_i + \boldsymbol{\beta}'\boldsymbol{x}_i^{(t)} + \gamma\, y_i^{(t-1)})}, \tag{3}$$

where $\Lambda(\cdot)$ is the cumulative distribution function (CDF) of the standard logit distribution. The above formulation implies estimating individual-specific intercepts α_i, $i = 1, \ldots, n$, which capture the effects on $Y_i^{(t)}$ of the unobserved individual-specific factors that are constant over time. These intercepts can be referred to as individual effects.

Estimation of the logit model parameters is typically straightforward and can be achieved by the maximum likelihood (ML) method. The ML estimator of $\boldsymbol{\theta} = \{\boldsymbol{\beta}, \gamma\}$ is obtained as follows:

$$\hat{\boldsymbol{\theta}} = \operatorname*{argmax}_{\boldsymbol{\theta}} \sum_{i=1}^{n}\sum_{t=1}^{T} \log p(Y_i^{(t)} = y_i^{(t)} | \boldsymbol{x}_i^{(t)}, y_i^{(t-1)}; \hat{\alpha}_i(\boldsymbol{\beta}, \gamma), \boldsymbol{\beta}, \gamma),$$

$$\hat{\alpha}_i(\boldsymbol{\beta}, \gamma) = \operatorname*{argmax}_{\alpha} \sum_{t=1}^{T} \log p(Y_i^{(t)} = y_i^{(t)} | \boldsymbol{x}_i^{(t)}, y_i^{(t-1)}; \alpha, \boldsymbol{\beta}, \gamma).$$

The above formulation refers to the dynamic fixed-effects logit case and can be easily applied to the other models discussed previously. For more details on the estimation procedures and algorithms for these models, see Aldrich & Nelson (1984) and Allison (2009).

3.2 Forecasting

Considering observed data for a sample of n units over T time occasions, it is appropriate to distinguish between:

- **in-sample prediction** at a time point $t \leq T$, whereby the model is estimated using all the available time occasions $\{1, \ldots, T\}$ and the occurrence of the event of interest for unit i at time t is predicted;
- **out-of-sample forecasting** at a future time point $t = t^*$, whereby the model is estimated using a restricted temporal window $\{t^* - \tau, \ldots, t^* - 1\}$ and the event of interest for unit i at time point t^* is forecast. Here τ denotes the size of the temporal window, corresponding to the number of time occasions used in the estimation phase.

In both cases, once the parameter estimates are obtained, the probability $p_i^{(t)}$ that the i-th individual experiences the event of interest at a given time t can be computed by substituting $\hat{\alpha}$ (or $\hat{\alpha}_i$), $\hat{\boldsymbol{\beta}}$, and $\hat{\gamma}$ along with the set of predictors $\boldsymbol{x}_i^{(t)}$ and the lagged response $y_i^{(t-1)}$ into Equations (1), (2), or (3). Afterward, an early warning for the event of interest is raised if the estimated probability $\hat{p}_i^{(t)}$ exceeds a threshold $c \in [0, 1)$. Adequate methods to choose c, such as those based on the ROC curve through Youden's J statistic, or the PR curve through the F1-score, have been discussed in Sect. 2.2.

4 Hidden Markov Models for Early Warning Systems

Since the pioneering work of Levinson et al. (1983), HM models (Bartolucci et al., 2013; Mor et al., 2021; Zucchini et al., 2016) have been widely used to forecast future observations; nowadays these models are commonly employed as EWS (Duprey & Klaus, 2022; Gupta et al., 2022; Jiang et al., 2016; Jouini et al., 2023). In the following section, we present the model formulation accounting for a binary response variable and covariates in the measurement model. We discuss maximum likelihood estimation and present both in-sample prediction and out-of-sample forecasting.

4.1 Model Formulation with Covariates

Considering the definition of the response variable $Y_i^{(t)}$ introduced in Sect. 3.1, let $\boldsymbol{Y}_i = (Y_i^{(1)}, \ldots, Y_i^{(T)})'$ denote the vector of responses for subject i. The HM model relies on individual-specific latent processes $\boldsymbol{U}_i = (U_i^{(1)}, \ldots, U_i^{(T)})'$, typically assumed to follow a first-order Markov chain with a finite number of latent states. These states, labeled from 1 to k, represent distinct subgroups within the heterogeneous population. Under the assumption of local independence, the response variables in $\boldsymbol{Y}_i$ are generally assumed to be conditionally independent given $\boldsymbol{U}_i$. In the following, we let $\boldsymbol{u}_i$ denote a realization of $\boldsymbol{U}_i$.

In the presence of individual covariates $\boldsymbol{X}_i^{(t)}$, defined as in Sect. 3.1, we assume that these covariates may directly influence the response variables. To relax the assumption of local independence, we also include the lagged response variable $Y_i^{(t-1)}$ among the covariates and we assume that the initial observation for every subject i, $Y_i^{(0)}$, is available. Including the lagged response among the covariates allows us to model the conditional serial dependence between observed responses over time (Bartolucci & Pennoni, 2007). In addition, we allow modeling unobserved heterogeneity between units, that is, the variability in individual responses that cannot be accounted for by observable covariates.

The HM model is formulated based on the following submodels:

(i) The **measurement submodel**, corresponding to $p(\boldsymbol{Y}_i = \boldsymbol{y}_i|\boldsymbol{U}_i = \boldsymbol{u}_i, \boldsymbol{X}_i = \boldsymbol{x}_i, Y_i^{(0)} = y_i^{(0)})$ and regarding the conditional distribution of the response vector $\boldsymbol{Y}_i$ given the latent process $\boldsymbol{U}_i$, the individual covariates in $\boldsymbol{X}_i$, and the initial observation $Y_i^{(0)}$. It relies on the conditional response probabilities, given the latent state and the covariate configuration, denoted as $\psi_{0|u\boldsymbol{x}_i^{(t)}y_i^{(t-1)}}$ and $\psi_{1|u\boldsymbol{x}_i^{(t)}y_i^{(t-1)}}$ for a binary response.

(ii) The **latent submodel**, corresponding to $p(\boldsymbol{U}_i = \boldsymbol{u}_i)$ and related to the distribution of the latent process. Its parameters are the initial probabilities, denoted as π_u, $u = 1, \ldots, k$, and subject to constraints $\pi_u \geq 0$ and $\sum_{u=1}^k \pi_u = 1$, and the transition probabilities, denoted as $\pi_{u|\bar{u}}$, $u, \bar{u} = 1, \ldots, k$, provided that $\pi_{u|\bar{u}} \geq 0$ and $\sum_{u=1}^k \pi_{u|\bar{u}} = 1$.

Concerning the measurement submodel, a parameterization in the form of a generalized linear model is employed to account for the effect of covariates (Bartolucci et al., 2013); considering that

$$\psi_{1|u\boldsymbol{x}_i^{(t)}y_i^{(t-1)}} = p(Y_i^{(t)} = 1|U_i^{(t)} = u, \boldsymbol{X}_i^{(t)} = \boldsymbol{x}_i^{(t)}, Y_i^{(t-1)} = y_i^{(t-1)}),$$

we have

$$\log \frac{\psi_{1|u\boldsymbol{x}_i^{(t)}y_i^{(t-1)}}}{\psi_{0|u\boldsymbol{x}_i^{(t)}y_i^{(t-1)}}} = \alpha_u + \boldsymbol{\beta}'\boldsymbol{x}_i^{(t)} + \gamma\, y_i^{(t-1)}, \tag{4}$$

for $t = 1, \ldots, T$ and $u = 1, \ldots, k$, where $\boldsymbol{\alpha} = (\alpha_1, \ldots, \alpha_k)'$ is the vector of support points of the latent variables, $\boldsymbol{\beta}$ is the vector of regression parameters for the covariates, and γ is the regression coefficient for the lagged response. The HM model may be seen as an extension of the models introduced in Sect. 3, which accounts for unobserved heterogeneity.

The manifest distribution of the response variables given the covariates is then expressed as follows:

$$p(\boldsymbol{Y}_i = \boldsymbol{y}_i | \boldsymbol{X}_i = \boldsymbol{x}_i, Y_i^{(0)} = y_i^{(0)}) = \\ = \sum_{\boldsymbol{u}} p(\boldsymbol{Y}_i = \boldsymbol{y}_i | \boldsymbol{U}_i = \boldsymbol{u}, \boldsymbol{X}_i = \boldsymbol{x}_i, Y_i^{(0)} = y_i^{(0)}) p(\boldsymbol{U}_i = \boldsymbol{u}),$$

where $\sum_{\boldsymbol{u}}$ denotes the sum over all possible configurations of latent variables. This distribution is computed by a forward recursion (Baum et al., 1970) as illustrated in Bartolucci et al. (2013, Chap. 3).

4.2 Maximum Likelihood Estimation

Given a sample of n subjects, the expectation–maximization (EM) algorithm (Dempster et al., 1977) is typically employed to maximize the log-likelihood function:

$$\ell(\boldsymbol{\theta}) = \sum_{i=1}^{n} \log p(\boldsymbol{Y}_i = \boldsymbol{y}_i | \boldsymbol{X}_i = \boldsymbol{x}_i, Y_i^{(0)} = y_i^{(0)})$$

thus performing ML estimation of the HM model parameters, collected in the vector $\boldsymbol{\theta}$. To this end, the EM algorithm relies on the complete-data log-likelihood function

$$\ell^*(\boldsymbol{\theta}) = \sum_{i=1}^{n} \log p(\boldsymbol{Y}_i = \boldsymbol{y}_i, \boldsymbol{U}_i = \boldsymbol{u}_i | \boldsymbol{X}_i = \boldsymbol{x}_i, Y_i^{(0)} = y_i^{(0)})$$

and, after a proper initialization of the model parameters, iteratively alternates the following two steps until convergence:

(i) **Expectation step**, which computes the conditional expected value of $\ell^*(\boldsymbol{\theta})$ given the observed data and the estimates of the model parameters at the previous step. This step is often computationally unfeasible and requires suitable recursions (see Welch (2003) for more details).
(ii) **Maximization step**, which updates the value of the parameters and maximizes the expectation of $\ell^*(\boldsymbol{\theta})$. In particular, explicit solutions exist for the initial and transition probabilities, while estimates for $\boldsymbol{\alpha}$ and $\boldsymbol{\beta}$ are obtained using numerical approaches such as the Newton–Raphson algorithm.

Convergence of the EM algorithm is evaluated on the basis of two common criteria, considering both the relative change in the log-likelihood function between two consecutive iterations and the corresponding change in the model parameter estimates. The algorithm reaches convergence and is stopped when

$$\frac{\ell(\boldsymbol{\theta}^{(h)}) - \ell(\boldsymbol{\theta}^{(h-1)})}{|\ell(\boldsymbol{\theta}^{(h-1)})|} < \varepsilon_1 \quad \text{and} \quad \max_s |\theta_s^{(h)} - \theta_s^{(h-1)}| < \varepsilon_2,$$

where ε_1 and ε_2 are suitable thresholds and h denotes the iteration number of the EM algorithm. It is also worth mentioning that the proposed HM-based EWS can be also formulated in a Bayesian framework, as shown in Tancini et al. (2024).

Initialization of the model parameters can be performed using both deterministic and random strategies. To address the potential convergence of the algorithm to local maxima of the log-likelihood function, multi-start strategies are commonly employed. Following this approach, the algorithm is executed multiple times with different initializations, and inference is based on the solution yielding the highest log-likelihood value. Other approaches are also available, such as those based on tempering (Brusa et al., 2023) and evolutionary algorithms (Brusa et al., 2024), which allow for a deeper inspection of the parameter space.

Given the estimated model, the conditional posterior distribution of the latent variables given the responses and the covariates is computed according to Bayes' theorem as follows:

$$q(\boldsymbol{U}_i = \boldsymbol{u}_i | \boldsymbol{X}_i = \boldsymbol{x}_i, Y_i^{(0)} = y_i^{(0)}, \boldsymbol{Y}_i = \boldsymbol{y}_i) = \\ = \frac{p(\boldsymbol{Y}_i = \boldsymbol{y}_i | \boldsymbol{U}_i = \boldsymbol{u}_i, \boldsymbol{X}_i = \boldsymbol{x}_i, Y_i^{(0)} = y_i^{(0)}) p(\boldsymbol{U}_i = \boldsymbol{u}_i)}{p(\boldsymbol{Y}_i = \boldsymbol{y}_i | \boldsymbol{X}_i = \boldsymbol{x}_i, Y_i^{(0)} = y_i^{(0)})}.$$

The predicted sequence of decoded states for each subject $i = 1, \ldots, n$ at each occasion t is obtained through local decoding. This maximum a posteriori procedure assigns individuals to a certain latent state at each time occasion t by maximizing each component of the posterior distribution:

$$\hat{u}_i^{(t)} = \underset{u=1,\ldots,k}{\text{argmax}}\ \hat{q}_{iu}^{(t)}, \qquad i = 1, \ldots, n,\ t = 1, \ldots, T,$$

with $\hat{q}_{iu}^{(t)} = q(U_i^{(t)} = u | \boldsymbol{X}_i = \boldsymbol{x}_i, Y_i^{(0)} = y_i^{(0)}, \boldsymbol{Y}_i = \boldsymbol{y}_i)$.

4.3 *In-Sample Prediction and Out-of-Sample Forecasting*

Given a sample of n units, in-sample prediction, defined in Sect. 3.2, relies on the estimated probability $\hat{p}_i^{(t)}$ of the binary event for unit i at time t. This is computed as the weighted average of the conditional response probabilities $\psi_{1|u\boldsymbol{x}_i^{(t)}}$ for each latent state u, with weights equal to the estimated posterior probabilities. In particular, relying on the parameterization expressed in Eq. (4), the conditional response probabilities may be estimated as

$$\hat{\psi}_{1|u\boldsymbol{x}_i^{(t)}y_i^{(t-1)}} = \frac{\exp(\hat{\alpha}_u + \hat{\boldsymbol{\beta}}'\boldsymbol{x}_i^{(t)} + \gamma\, y_i^{(t-1)})}{1 + \exp(\hat{\alpha}_u + \hat{\boldsymbol{\beta}}'\boldsymbol{x}_i^{(t)} + \gamma\, y_i^{(t-1)})}, \tag{5}$$

for latent state $u = 1, \ldots, k$ and any possible covariate configuration $\boldsymbol{x}_i^{(t)}$, which may also include the lagged response variable. As for logit and probit models, an early warning signal for the event of interest may be identified when the probability $\hat{p}_i^{(t)}$ exceeds a suitable threshold $c \in [0, 1)$ possibly chosen through the ROC or the PR analysis.

In-Sample Prediction. This procedure is based on the estimated posterior probabilities of the latent variable $U_i^{(t)}$ given the responses $\boldsymbol{y}_i$, the covariates $\boldsymbol{x}_i$, and the initial response $y_i^{(0)}$, denoted by $\hat{q}_{iu}^{(t)}$. The probability $p_i^{(t)}$ of the event of interest for unit i at time t is then computed as follows:

$$\hat{p}_i^{(t)} = \sum_{u=1}^{k} \hat{q}_{iu}^{(t)}\, \hat{\psi}_{1|u\boldsymbol{x}_i^{(t)}y_i^{(t-1)}}.$$

Out-Of-Sample Forecasting. Forecasting the response variable in a time period immediately following the observed one using an HM model involves determining the forecasting distribution; see also Bartolucci et al. (2013). This procedure is based on the estimated conditional probabilities of the latent variable $U_i^{(t^*)}$ at a future time point t^* given the responses $\boldsymbol{y}_i$, the covariates $\boldsymbol{x}_i$, and the initial response $y_i^{(0)}$:

$$\hat{q}_{iu}^{(t^*)} = \sum_{\bar{u}=1}^{k} \hat{\pi}_{u|\bar{u}}\, \hat{q}_{i\bar{u}}^{(t^*-1)}.$$

The probability $p_i^{(t^*)}$ of the event of interest for unit i at time t^* is then estimated, similarly to the previous case, as

$$\hat{p}_i^{(t^*)} = \sum_{u=1}^{k} \hat{q}_{iu}^{(t^*)}\, \hat{\psi}_{1|u\boldsymbol{x}_i^{(t^*)}y_i^{(t^*-1)}}.$$

5 Application: Early Warnings of Financial Crises

In the following, we apply both traditional prediction methods and the proposed approach based on HM models to develop EWSs. Specifically, in Sect. 5.1, we describe the data mentioned in Sect. 1. In Sect. 5.2, we show how to perform early

Table 2 Observed crises of the 129 countries over the period from 1983 to 2017

Variable	Abs. value	Rel. frequency
Crisis	230	0.069
Lagged crisis	234	0.073

prediction of crises using dynamic logit and probit models under pooled and fixed-effects specifications. Finally, in Sect. 5.3, we present the results obtained with the HM model with covariates and the lagged response variable and we evaluate the resulting in-sample prediction and out-of-sample forecasting.

5.1 Prediction of Countries' Financial Crises

The data, retrieved from Pigini (2021), cover the period from 1983 to 2017 and are referred to 129 countries. Crisis records, as well as the 1-year lagged variable, are illustrated in Laeven & Valencia (2018); the response variable refers to systemic banking crises as defined in Sect. 1. The original number of crises (219) has been increased based on the information provided by the lagged variable, resulting in a total of 230 crisis events, with an incidence of 6.9% over the whole period. Table 2 provides detailed descriptive statistics for both crisis and lagged crisis variables. Macroeconomic, monetary, and financial indicators are obtained from the International Financial Statistics by the International Monetary Fund (https://www.imf.org) and the World Development Indicators by the World Bank (https://databank.worldbank.org). They include the real GDP growth rate, logarithm of per capita GDP, inflation rate, real interest rate, broad money to foreign exchange reserves ratio, growth of private credit, and growth rate of net foreign assets to GDP. We refer to Chap. Sampling-Based and Cost-Sensitive Classification in Early Warning Systems for Financial Crises for a more detailed description of these variables and for the list of countries experiencing crises. Partially missing values in the covariates are imputed as 0 and accounted for using dummy variables that act as missingness indicators. This approach allows us to retain all observations in the analysis while evaluating the informativeness of covariates with missing observations; see, among others, Dardanoni et al. (2011). Summary statistics for such variables are reported in Table 3.

5.2 Results of the Binary Regression Models: Out-of-Sample Prediction

In the following, we show the out-of-sample forecasting performance of the logit, probit, and logit fixed-effects models, including all the covariates and the lagged

Table 3 Descriptive statistics of predictors related to 129 countries observed over the period from 1983 to 2017

Variable	Mean	Std. dev.	Min.	Max.
Real GDP growth	3.88	4.24	−13.82	18.18
Log per capita GDP	7.71	1.46	4.71	11.39
Inflation	26.50	324.74	−31.57	13611.64
Real interest rate	2.38	11.23	−46.11	66.15
M2 to foreign exchange reserves	15.21	33.58	0.03	145.00
Growth of real domestic credit	16.25	32.86	−631.04	518.08
Growth of net foreign assets to GDP	0.02	0.23	−2.44	2.30

response. In particular, to predict banking crises in year 2003, models are estimated using data from 1983 to 2002; for the prediction of crises in 2004, models are re-estimated, including observations from 1984 to 2003. This rolling window procedure is repeated annually to provide predictions for each year in the range 2003–2017. For each of these year, we evaluate model performance using both ROC and PR analyses. Values of AUROC and AUPR, the corresponding optimal thresholds $\tilde{c}$, and the number of predicted crises and false alarms are reported for each case.

Table 4 provides a summary of forecasting performance over the entire evaluation period. The results indicate that the logit model has the best overall performance in terms of total predicted crises, irrespective of the evaluation metric adopted.

Considering the threshold based on the F1-score, the logit and probit models correctly predict 37 out of 49 crisis events (with 14 false alarms), while the fixed-effects model only forecasts 32 crises (with 9 false alarms). On the contrary, Youden's index-based selection threshold correctly identifies 37, 38, and 31 crises with logit, probit, and logit fixed-effects models, respectively, but reveals a considerably higher number of false alarms, ranging from 47 to 78 across the models. Furthermore, approaches based on PR and ROC curves have similar performance for the three models, but the PR-based metrics provide a more balanced assessment by better accounting for the incidence of false alarms. Overall, these results suggest that the logit and probit models have comparable predictive performance, whereas the logit fixed-effects model shows some limitations in detecting crisis events, notwithstanding the method used to select the threshold.

Regarding the fixed-effects model, while in theory accounting for individual heterogeneity could improve predictive accuracy, this benefit vanishes when the positive class, namely the crisis event, is rare, like in this case. In particular, for countries not experiencing any crisis event over the estimation period, the country-specific

Table 4 Number of predicted crises, false alarms, and main performance metrics for the out-of-sample forecasting for logit, probit, and logit fixed-effects models for the years from 2003 to 2017 when a total number of 49 financial crises is observed; results in the upper panel are related to the Precision–Recall (PR) curve and those in the lower panel to the Receiver Operating Characteristic (ROC) curve

	Logit	Probit	Fixed effects
	PR curve		
Predicted crises	37	37	32
False alarms	14	14	9
Sensitivity/Recall	0.755	0.755	0.653
Specificity	0.993	0.993	0.991
Precision	0.725	0.725	0.780
Accuracy	0.987	0.987	0.974
F1-score	0.740	0.740	0.711
	ROC curve		
Predicted crises	37	38	31
False alarms	47	52	78
Sensitivity/Recall	0.776	0.776	0.776
Specificity	0.938	0.938	0.934
Precision	0.245	0.247	0.376
Accuracy	0.934	0.934	0.926
F1-score	0.373	0.374	0.507

intercept tends to diverge toward infinity and does not contribute to the estimation process. In this case, this problem reduces the predictive performance of the model. This is a well-known problem in the statistical literature, commonly referred to as the separation problem (Albert & Anderson, 1984; Heinze & Schemper, 2002).

5.3 *Results of the Hidden Markov Model*

In the following, we present the results of the HM model applied to the data described in Sect. 5.1. We first focus on model selection and estimation, analyzing the estimated parameters and interpreting the latent states. Then, we assess the approach in terms of in-sample prediction and out-of-sample forecasting.

Model Estimation. The HM model introduced in Sect. 4 is estimated using data on the occurrence of banking crises from 1983 to 2017 and incorporating all the related covariates, as well as the lagged response variable. To determine the optimal number of latent states, the model is estimated for a number of states k ranging from 1 to 6. Selection of the optimal number is based on the Akaike information criterion (AIC, Akaike, 1974), which is one of the most commonly used criteria,

Table 5 Maximum log-likelihood function at convergence ($\hat{\ell}$), number of parameters (#par), and AIC value resulting from fitting the HM model introduced in Sect. 4 for values of k ranging from 1 to 6; values in bold refer to the best result

k	$\hat{\ell}$	#par	AIC
1	−522.076	12	1068.152
2	−515.173	15	1060.346
3	**−505.861**	**20**	**1051.722**
4	−500.120	27	1054.239
5	−494.616	36	1061.232
6	−493.411	47	1080.821

along with the Bayesian information criterion (Schwarz, 1978). For an evaluation of the performance of these indices in the context of HM model selection, see Bacci et al. (2014). Each model is estimated 20 times using a multi-start strategy that combines deterministic and random starting rules. The estimate of the parameter vector $\hat{\boldsymbol{\theta}}$ is then taken as the one corresponding to the highest likelihood value.

Results of the model selection procedure are reported in Table 5, which indicates that the model with three hidden states provides the best compromise between model fit and parsimony it reaches a maximum log-likelihood equal to −505.861 with 20 parameters and AIC equal to 1051.722.

Model Results. The adopted model identifies three subpopulations of countries with different levels of financial vulnerability. These groups are ordered according to the estimated support points $\hat{\alpha}_u$ of the latent variable, as in Eq. (4). The highest estimated value corresponds to the group of countries experiencing a high financial distress regime (first hidden state), with $\hat{\alpha}_1 = 87.753$, followed by countries facing low financial challenges (second hidden state), with $\hat{\alpha}_2 = 12.167$, and those experiencing negligible financial distress (third hidden state), with $\hat{\alpha}_3 = -34.966$. The states are represented in the path diagram of Fig. 1 and are colored in red, yellow, and green for the first, second, and third state, respectively. The estimated initial probabilities are displayed within the nodes. It can be observed that, at the beginning of the observation period in 1983, the first group includes approximately 9% of the countries, the second 50%, and the third 41%.

To better characterize these states, Fig. 2 displays the estimates of the average conditional probability of financial crisis for each group obtained as follows:

$$\bar{\phi}_{1|u}^{(t)} = \frac{1}{n} \sum_{i=1}^{n} \hat{\phi}_{1|u\boldsymbol{x}_i^{(t)} y_i^{(t-1)}},$$

where $\hat{\phi}_{1|u\boldsymbol{x}_i^{(t)} y_i^{(t-1)}}$ is estimated according to Eq. (5). We notice that countries in the first subpopulation exhibit a high probability of experiencing an important financial distress throughout the considered years, with estimates ranging from 0.275 in 2017

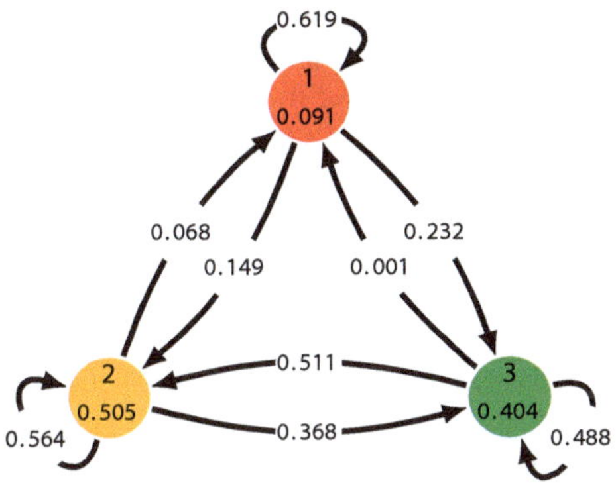

Fig. 1 Transition graph and estimated parameters of the HM latent submodel with three hidden states

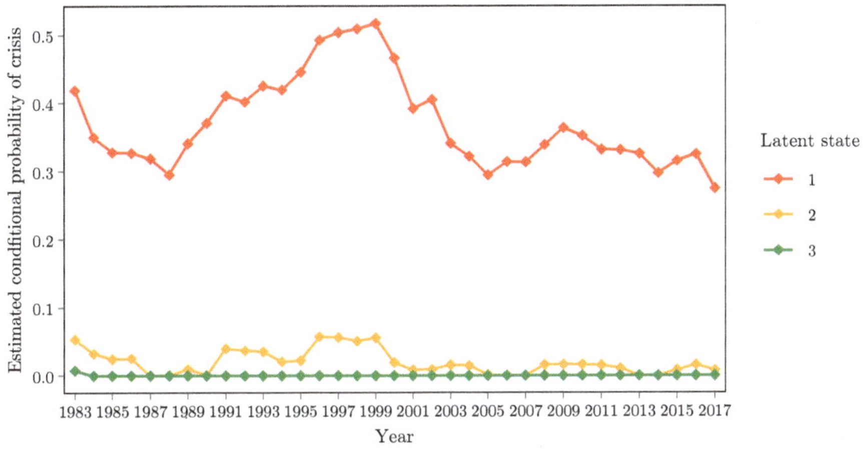

Fig. 2 Average conditional probability of financial crises $\bar{\phi}_{1|u}^{(t)}$ for each hidden state u at every time occasion t under the estimated HM model with three hidden states

to 0.517 in 1999. The highest probabilities are observed during the period 1996–1999 (Asian financial crisis),[1] and in specific years such as 1983 (Mexican debt crisis)[2], 2002 (end of dot-com bubble)[3], and 2009 (failure of Lehman Brother).[4] Countries in the second group have significantly lower probabilities of suffering from financial

[1] For more details, see https://en.wikipedia.org/wiki/1997_Asian_financial_crisis, accessed in April 2025.

[2] For more information, see https://en.wikipedia.org/wiki/Latin_American_debt_crisis, accessed in April 2025.

[3] For more details, see https://en.wikipedia.org/wiki/Dot-com_bubble, accessed in April 2025.

[4] For more details, see https://en.wikipedia.org/wiki/Bankruptcy_of_Lehman_Brothers, accessed in April 2025.

distress and exhibit fewer jumps and less variability over time compared to the first group. The estimated probabilities of experiencing crisis episodes vary over time, with the highest value of 0.057 observed in 1996 during the Asian financial crisis, and the lowest, 0.000, recorded in 2014. As for the third group, financial distress levels are remarkably low, with the highest value being equal to 0.008 in 1983, with minimal fluctuations in estimated probabilities over time.

The estimated average transition probabilities among states are reported next to the corresponding arrows in the transition diagram shown in Fig. 1. We observe that countries in the second hidden state (yellow), characterized by low financial distress, have a probability of approximately 0.068 of increasing their distress level by transitioning to the first group (red). Countries in the third hidden state (green), characterized by a negligible risk of financial crises, show an estimated probability of 0.511 of transitioning to the second group (yellow), thereby increasing their risk during the period. Finally, countries in the high-risk state (red) have a 0.149 probability of moving to the low-risk state (yellow), while the probability of transitioning from the low risk to the negligible-risk group (green) is 0.368.

As illustrated in Sect. 4.2, a key feature of the HM model is the estimated most probable sequence of hidden states given the observed longitudinal data. Using local decoding, this sequence is determined for each country based on the highest estimated posterior membership probability at each time occasion. Figure 3 depicts the decoded sequence for each country with different colors referred to each state. Throughout the analyzed time period, approximately 79.1–99.2% of the countries are classified in the second group (low financial distress). The proportions of countries classified into the other two groups are significantly lower, ranging from 0% to 15.5% for the first group (high financial distress) and from 0% to 6.2% for the third group (negligible financial distress). It is observed that almost all countries are classified in the high financial distress state at least once during the analyzed period. Moreover, note that the high financial distress phase is typically preceded by a period of mild financial distress and followed by a crisis-free phase, which, however, is of relatively short duration. The graph allows us to examine the sequence of decoded latent states for specific countries, illustrating, for example, the behavior of Asian nations during the Asian financial crisis from 1997 to 2001. Only three countries shown in the lower part of the graph—Ukraine, Moldova, and Guinea-Bissau—are classified as high risk in recent years. Notably, Morocco is the only country that is never classified in the highest risk state. Countries located in the lower part of the graph, such as Panama, experienced high financial distress mainly in the early 1980s. Some countries, like Japan and Indonesia, were affected by high financial risk in the second half of the 1990s through the early 2000s. The group of countries in the upper part of the graph, including the United States and Iceland, were most impacted by the 2007 financial crisis. These countries are in fact classified as high risk for several consecutive years, up until 2013.

A world map colored according to the decoded states for years 1995 and 1999 is also provided for comparison, in Fig. 4 in Appendix A.1. Moreover, Table 6 shows the estimated regression coefficients from Eq. (4), along with the corresponding standard errors and p-values.

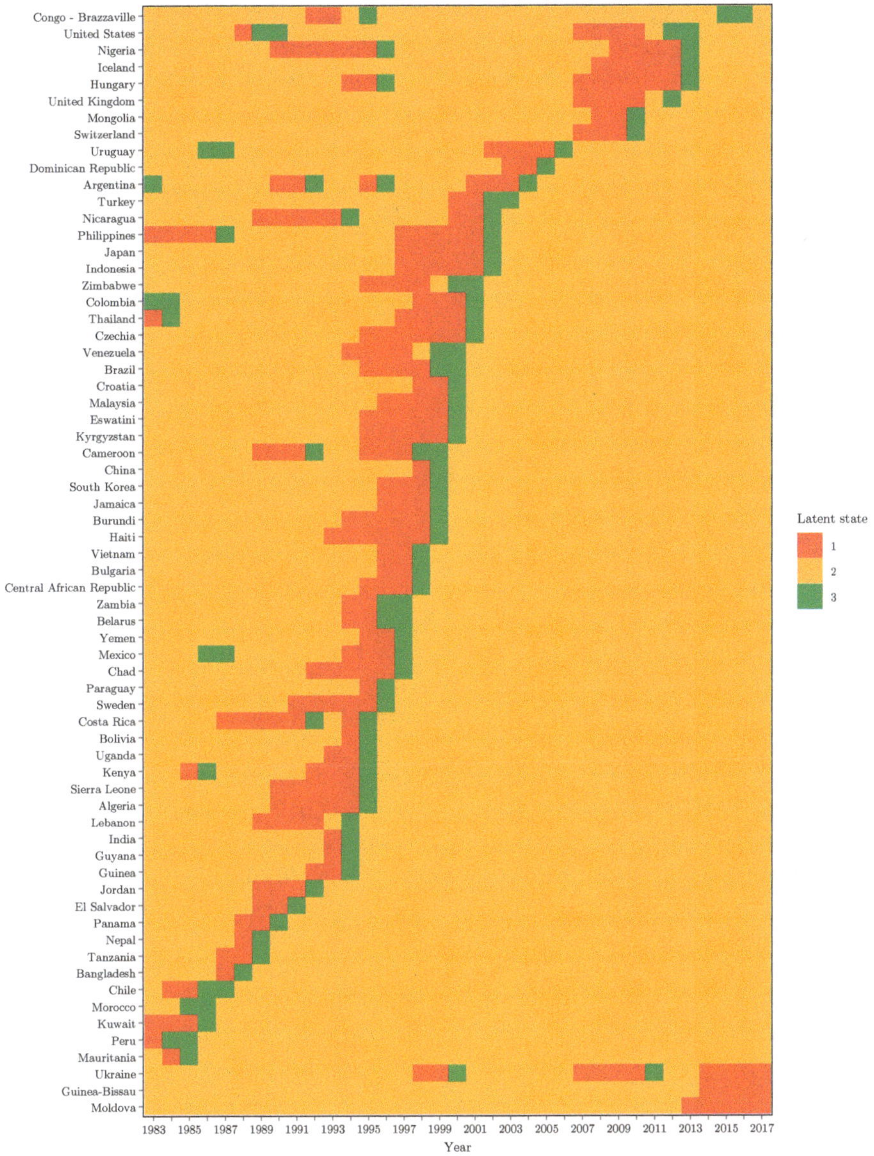

Fig. 3 Sequence of locally decoded states for each country and year under the estimated HM model with three hidden states

Table 6 Estimates of the regression parameters affecting the conditional response probabilities, along with their standard errors and corresponding p-values under the estimated HM model with three hidden states. Significance is indicated at the †10% and *5% levels

Covariate	$\hat{\beta}$	$s.e.$	p-value
Lagged crisis	$71.316^{\dagger}$	42.861	0.096
Growth of real GDP	−0.298*	0.140	0.033
Log per capita GDP	−0.190	0.266	0.474
Inflation	−0.026	0.032	0.415
Real interest rate	0.366*	0.152	0.016
M2 to foreign exchange reserves	0.147*	0.064	0.021
Growth of real domestic credit	0.341*	0.138	0.013
Growth of net foreign assets to GDP	−32.315*	12.855	0.012
Dummy1	−3.503	2.272	0.123
Dummy2	4.688*	2.024	0.021

These coefficients serve as crisis signals and represent early warning indicators. Almost all regression parameters are statistically significant, except for inflation and the logarithm of per capita GDP. The negative sign of some coefficients indicates that, for a given country in a specific latent state, the conditional probability of a financial crisis decreases as the corresponding indicator increases. For example, the exponential of the estimated coefficient for real GDP growth is $\exp(-0.298) = 0.742$, representing the odds ratio of experiencing a financial crisis associated with a unit increase in real GDP growth, while holding the other covariates constant. This means that the odds of experiencing a banking crisis decreases by approximately 26% for a one-unit increase in real GDP growth, at fixed values of the other covariates. Conversely, a positive coefficient, such as that of the real interest rate (0.366), suggests that an increase of one percentage point in the real interest rate raises the odds in favor of a crisis by 1.44 compared to a non-crisis scenario.

In-Sample Prediction. In-sample predictions are obtained by computing the predicted probabilities of financial crises, $\hat{p}_i^{(t)}$, for each country i and year t, as outlined in Sect. 4.3. Table 7 presents the performance of the model in terms of correctly predicted crises, false alarms, and evaluation metrics.

The threshold selected using the PR curve is higher than that chosen with the ROC curve (0.19 versus 0.08). Consequently, the PR-based approach predicts fewer financial crises, correctly identifying 211 out of 230 cases (sensitivity: 0.917), compared to the ROC-based approach, which recognizes all crises (sensitivity: 1.000). On the other hand, the approach based on the PR curve generates fewer false positives, with only two instances erroneously predicted as crises, corresponding to a specificity approximately equal to 1.000; the ROC-based method produces 35 false alarms,

Table 7 Optimal threshold, number of predicted crises and false alarms, and main performance metrics for the in-sample prediction under the HM model with three hidden states for the years from 2003 to 2017 when a total number of 49 financial crises is observed. Results on the left are related to Precision–Recall (PR) curve and those on the right to the Receiver Operating Characteristic (ROC) curve

	PR curve	ROC curve
Threshold ($\tilde{c}$)	0.190	0.080
Predicted crises	211	230
False alarms	2	35
Sensitivity/Recall	0.917	1.000
Specificity	1.000	0.992
Precision	0.991	0.868
Accuracy	0.995	0.992
F1-score	0.953	0.929

leading to a specificity of 0.992. Overall, the two approaches yield very similar and high accuracy values, reflecting the proportion of all correct classifications: 0.995 for the PR-based method and 0.992 for the ROC-based approach. Additional details on the predicted crises are provided in Appendix A.2, where we compare predictions for models whose number of hidden states increases from 1 to 6. As shown in Table 10, in the appendix, increasing the number of states results in exact in-sample predictive accuracy.

Out-Of-Sample Forecasting. Generating predictions for the response variable in the period immediately following the one observed in the HM model involves determining the forecasting distribution; see also Bartolucci et al. (2013). As illustrated in Sect. 4.3, out-of-sample forecasts require the following four steps to be performed, where the years to be predicted are in sequence $t^* = 2003, \ldots, 2017$:

1. The data are divided into a training set and a test set. The training set includes observations from all countries for the years within the temporal window $\{t^* - 20, \ldots, t^* - 1\}$, while the test set is restricted to the year t^* to be predicted.
2. The HM model is estimated on the training data, considering a number of latent states k ranging from 1 to 4.
3. In-sample prediction is performed for each of the estimated models varying the number of hidden states k and calculating the F1-score for each value of k; then, the model with the highest F1-score is selected as the best one, along with its corresponding optimal threshold $\tilde{c}$.
4. On the basis of the estimated parameters of the selected model, we perform out-of-sample forecasting, using $\tilde{c}$ as the threshold for prediction; see Sect. 2.2.

The evaluation metrics resulting from this procedure are shown in Table 8. Despite the imbalanced response variable, the accuracy rate indicates that 98.7% of observations are correctly classified. In particular, the value of sensitivity is 0.735, highlighting that the proposed approach correctly predicts 73.5% of financial crises occurring

Table 8 Number of predicted crises and false alarms, and main performance metrics for the out-of-sample forecasting under the HM model with three hidden states for the years from 2003 to 2017 when a total number of 49 financial crises is observed; results are related to Precision–Recall (PR) curve

	PR curve
Predicted crises	36
False alarms	13
Sensitivity/Recall	0.735
Specificity	0.993
Accuracy	0.987
Precision	0.735
F1-score	0.735

Table 9 Number of correctly predicted crises and false alarms obtained with the out-of-sample forecasting under the proposed HM model approach for the years from 2003 to 2017

Year	Total	Predicted (%)	False alarms
2003	3	2 (66.67)	0
2004	2	2 (100.00)	2
2005	1	1 (100.00)	1
2006	0	0 (–)	1
2007	2	0 (0.00)	0
2008	7	2 (28.57)	0
2009	8	7 (87.50)	0
2010	6	6 (100.00)	2
2011	5	4 (80.00)	0
2012	3	3 (100.00)	1
2013	0	0 (–)	3
2014	3	0 (0.00)	0
2015	3	3 (100.00)	1
2016	3	3 (100.00)	2
2017	3	3 (100.00)	0
Total	49	36 (74.06)	13

between 2003 and 2017 (36 out of 49). The number of false alarms is particularly low (13), as reflected by the specificity value of 0.993. Since similar results are observed when using the Youden's J statistic to select the threshold instead of the F1-score, these results are not reported in Table 8. We only mention that with this measure the model forecasts the same number of crises and 28 false alarms.

Table 9 presents the number of correctly forecasted crises and false alarms in each year. Due to the first-order assumption of the hidden chain, the HM model provides a forecasting distribution that is much more accurate in predicting a crisis that occurs immediately after the previous one. However, this forecasting procedure is less accurate in detecting crises that are further apart in time. An additional analysis of the forecasting performance of the HM model is provided in Appendix A.3, where we examine alternative model formulations. In that appendix, we compare the ROC curves of HM models using a temporal window of 10 or 20 years, with and without the inclusion of the lagged response variable. Our analysis show that using a larger temporal windows enhances forecasting performance, and that including the lagged response as covariate leads to a substantial increase in predictive accuracy.

6 Conclusions

Short-term forecasts have recently become increasingly widespread across many fields, as they are essential for planning and decision-making. However, providing reliable predictions of the future remains a significant challenge. In particular, there is a growing demand for probabilistic forecasts based on methods that are accessible, reproducible, and interpretable. In this chapter, we present probabilistic prediction and forecasting approaches using simple and more complex statistical models tailored to the analysis of cross-sectional and longitudinal data.

Regression models for binary response have been widely used as early warning systems to forecast events. As shown by the analysis of financial crisis data in various countries, binary regression models offer reliable overall forecasting performance. Moreover, they constitute a simple approach: easy to implement, fast to compute, and straightforward to interpret for practitioners.

The hidden Markov (HM) model represents a more complex yet robust alternative for early warning systems compared to these traditional approaches. Although it may yield slightly lower out-of-sample performance, this approach offers several advantages: it enables classification of units into distinct risk-level groups, estimation of covariate effects on the probability of adverse events, thereby providing multi-hazards early warnings, and assessment of serial dependence. Furthermore, it supports the monitoring of units over time through time-varying classification of such units into different risk levels. The model can be updated in real time to reflect the evolving dynamics of the phenomenon under study.

In conclusion, both models provide accurate and relatively straightforward prediction and forecast with interpretable estimates that produce useful early warning signals. Logit regression models may be suggested when the primary goal is prediction. However, when the analysis requires accounting for the longitudinal nature of the data and identifying (dynamic) latent subpopulations, the HM model provides a more complete framework.

It should also be noted that the out-of-sample results are obtained using an imbalanced training and test set. A potential area for further investigation is to evaluate the performance of the proposed models when applying more sophisticated balancing techniques to both the training and test datasets. Additional research will also explore potential non-linear associations and interactions among key indicators, as well as the use of penalized likelihood methods, which could simplify the estimation procedure and lead to more accurate forecasts.

Appendix A: Additional Results of the Hidden Markov Model

In the following, we provide additional results related to the HM model. In particular, Appendix A.1 illustrates the world map obtained by assigning each country to a latent state through local decoding. Appendix A.2 outlines the in-sample prediction performance of the HM model across a varying number of latent components. Finally, A.3 examines the out-of-sample forecasting performance for different configurations of the HM model.

A.1 Decoding

To better illustrate the results of the proposed HM model formulation, we provide world maps for the years 1995 and 1999, where each country is colored according to the a posteriori classification. The plot, shown in Fig. 4, illustrates differences between countries and compares them immediately before (1995) and during the Asian financial crisis. It highlights that, in 1995, Asian countries were allocated to the second latent state, whereas in 1999 many of them, including Indonesia, Japan, Malaysia, the Philippines, and Thailand, were assigned to the first latent state, characterized by the highest level of financial distress.

A.2 In-sample Prediction Results

In order to compare the performance of the HM model, we propose evaluating the main metrics as the number of latent states varies from 1 to 6. It is worth noting that the model selection presented in Sect. 5.3 was based on the AIC. Table 7 shows the main metrics discussed in the paper (see Sect. 2.2), obtained for $k = 1, \ldots, 6$. The results shown in the table indicate that, when the number of latent states increases (e.g., 4, 5, and 6), the model correctly predicts all financial crises without generating false alarms.

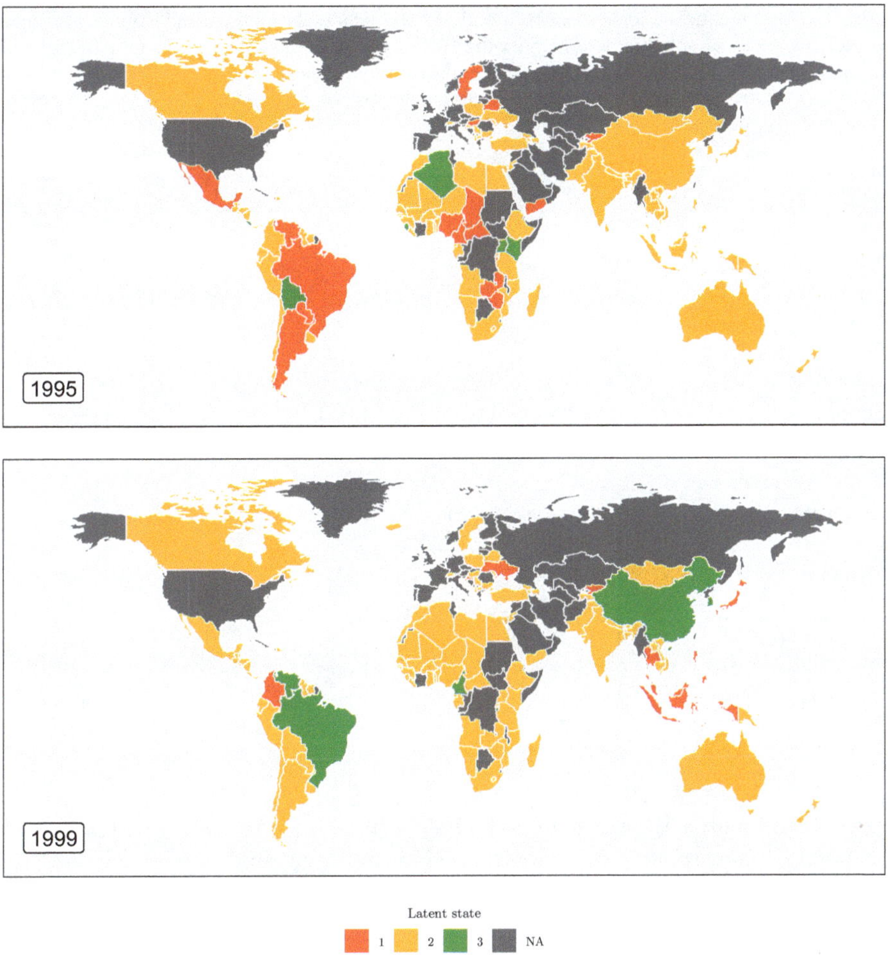

Fig. 4 World maps colored according to the classification of each country obtained for the years 1995 (upper panel) and 1999 (bottom panel) under the HM model with three latent states; countries shaded in gray are those for which information on financial crises is not available

A.3 Out-of-Sample Forecasting Results

We also compare the out-of-sample forecasting performance of the HM model across different formulations in which the number of latent states is fixed to three. In particular, we evaluate the model by varying the size of the temporal window $\{t^* - \tau, \ldots, t^* - 1\}$ used for the training set (10 or 20 years), and by including or excluding the lagged response variable from the set of covariates. The results presented in Fig. 5 highlight the importance of accounting for the serial dependence, as the models that include the lagged response achieve the highest area under the ROC

Table 10 Optimal thresholds, number of predicted crises and false alarms, and main performance metrics for the in-sample prediction under the HM model with hidden states ranging from 1 to 6 for the years from 2003 to 2017 when a total number of 49 financial crises are observed; results on the top are related to Precision–Recall (PR) curve and those on the right to the Receiver Operating Characteristic (ROC) curve

	$k = 1$	$k = 2$	$k = 3$	$k = 4$	$k = 5$	$k = 6$
AIC	1068.152	1060.346	1051.722	1054.239	1061.232	1080.821
	ROC curve					
Threshold ($\tilde{c}$)	0.030	0.060	0.080	0.150	0.370	0.100
Predicted crises	173	230	230	230	230	230
False alarms	397	28	35	0	0	0
Sensitivity	0.752	1.000	1.000	1.000	1.000	1.000
Specificity	0.907	0.993	0.992	1.000	1.000	1.000
	PR curve					
Threshold ($\tilde{c}$)	0.210	0.240	0.190	0.150	0.370	0.100
Predicted crises	153	210	211	230	230	230
False alarms	81	0	2	0	0	0
Sensitivity	0.665	0.913	0.917	1.000	1.000	1.000
Specificity	0.981	1.000	1.000	1.000	1.000	1.000

curve: 0.889 with a 10-year temporal window (blue line) and 0.923 with a 20-year temporal window (pink line). In contrast, when the lagged response is excluded from the model, this value drops to 0.436 with a 10-year temporal window (brown line).

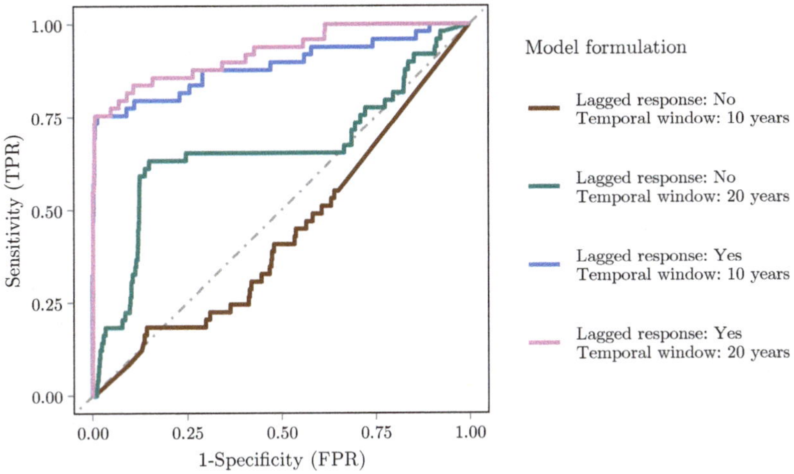

Fig. 5 Area under the ROC curve for different specifications of the HM model with three latent states, considering temporal windows of 20 or 10 years, with and without the inclusion of the lagged response variable

References

Agresti, A. (2013). *Categorical data analysis* (3rd ed.). Hoboken: Wiley.

Akaike, H. (1974). A new look at the statistical model identification. *IEEE Transactions on Automatic Control, 19*, 716–723.

Albert, A., & Anderson, J. A. (1984). On the existence of maximum likelihood estimates in logistic regression models. *Biometrika, 71*, 1–10.

Aldrich, J. H., & Nelson, F. D. (1984). *Linear probability, logit, and probit models*. Newbury Park, CA: Sage Publications.

Allison, P. D. (2009). *Fixed effects regression models*. Thousand Oaks, CA: SAGE Publications.

Antunes, A., Bonfim, D., Monteiro, N., & Rodrigues, P. M. (2018). Forecasting banking crises with dynamic panel probit models. *International Journal of Forecasting, 34*, 249–275.

Bacci, S., Pandolfi, S., & Pennoni, F. (2014). A comparison of some criteria for states selection in the latent Markov model for longitudinal data. *Advances in Data Analysis and Classification, 8*, 125–145.

Bartolucci, F., Farcomeni, A., & Pennoni, F. (2013). *Latent markov models for longitudinal data*. Boca Raton: Chapman and Hall/CRC.

Bartolucci, F., Pandolfi, S., & Pennoni, F. (2022). Discrete latent variable models. *Annual Review of Statistics and Its Application, 9*, 425–452.

Bartolucci, F., & Pennoni, F. (2007). A class of latent Markov models for capture-recapture data allowing for time, heterogeneity, and behavior effects. *Biometrics, 63*, 568–578.

Basher, R. (2006). Global early warning systems for natural hazards: systematic and people-centred. *Philosophical Transactions of the Royal Society A: Mathematical, Physical and Engineering Sciences, 364*, 2167–2182.

Baum, L. E., Petrie, T., Soules, G., & Weiss, N. (1970). A maximization technique occurring in the statistical analysis of probabilistic functions of Markov chains. *The Annals of Mathematical Statistics, 41*, 164–171.

Berrar, D., & Flach, P. (2012). Caveats and pitfalls of ROC analysis in clinical microarray research (and how to avoid them). *Briefings in Bioinformatics, 13*, 83–97.

Bharati, M., & Lomte, S. (2016). A survey on hidden Markov model (HMM) based intention prediction techniques. *International Journal of Engineering Research and Applications, 6*, 167–172.

Brusa, L., Bartolucci, F., & Pennoni, F. (2023). Tempered expectation-maximization algorithm for the estimation of discrete latent variable models. *Computational Statistics, 38*, 1391–1424.

Brusa, L., Pennoni, F., & Bartolucci, F. (2024). Maximum likelihood estimation for discrete latent variable models via evolutionary algorithms. *Statistics and Computing, 34*, 1–15.

Caggiano, G., Calice, P., Leonida, L., & Kapetanios, G. (2016). Comparing logit-based early warning systems: Does the duration of systemic banking crises matter? *Journal of Empirical finance, 37*, 104–116.

Chen, P., Liu, R., Li, Y., & Chen, L. (2016). Detecting critical state before phase transition of complex biological systems by hidden Markov model. *Bioinformatics, 32*, 2143–2150.

Dardanoni, V., Modica, S., & Peracchi, F. (2011). Regression with imputed covariates: A generalized missing-indicator approach. *Journal of Econometrics, 162*, 362–368.

Dasgupta, S., & Wheeler, D. R. (2020). Modeling and Predicting the Spread of Covid-19: Comparative Results for the United States, the Philippines, and South Africa. Policy Research Working Paper, 9419, The World Bank.

David, R., & Söffker, D. (2023). A review on machine learning-based models for lane-changing behavior prediction and recognition. *Frontiers in Future Transportation, 4*, Article 950429.

Davis, E. P., & Karim, D. (2008). Comparing early warning systems for banking crises. *Journal of Financial stability, 4*, 89–120.

Davis, J., & Goadrich, M. (2006). The relationship between Precision-Recall and ROC curves. *Proceedings of the 23rd International Conference on Machine Learning* (pp. 233–240). New York: Association for Computing Machinery.

Demirgüç-Kunt, A., & Detragiache, E. (2005). Cross-country empirical studies of systemic bank distress: a survey. *National Institute Economic Review, 192*, 68–83.

Dempster, A. P., Laird, N. M., & Rubin, D. B. (1977). Maximum likelihood from incomplete data via the EM algorithm. *Journal of the Royal Statistical Society: Series B, 39*, 1–22.

Duprey, T., & Klaus, B. (2022). Early warning or too late? A (pseudo-)real-time identification of leading indicators of financial stress. *Journal of Banking & Finance, 138*, 1–20.

Fawcett, T. (2006). An introduction to ROC analysis. *Pattern Recognition Letters, 27*, 861–874.

Gregoir, S., & Lenglart, F. (2000). Measuring the probability of a business cycle turning point by using a multivariate qualitative hidden Markov model. *Journal of Forecasting, 19*, 81–102.

Gupta, A., Garg, D., & Kumar, P. (2022). Mining sequential learning trajectories with hidden Markov models for early prediction of at-risk students in e-Learning environments. *IEEE Transactions on Learning Technologies, 15*, 783–797.

Hanley, J. A., & McNeil, B. J. (1982). The meaning and use of the area under a receiver operating characteristic (ROC) curve. *Radiology, 143*, 29–36.

Heckman, J. J. (1981). Heterogeneity and state dependence. In S. Rosen (Ed.), *Studies in Labor Markets* (pp. 91–140). Chicago: University of Chicago Press.

Heinze, G., & Schemper, M. (2002). A solution to the problem of separation in logistic regression. *Statistics in Medicine, 21*, 2409–2419.

Hsiao, C. (2022). *Analysis of Panel Data*. New York: Cambridge University Press.

Jiang, P., Liu, X., Zhang, J., & Yuan, X. (2016). A framework based on hidden Markov model with adaptive weighting for microcystin forecasting and early-warning. *Decision Support Systems, 84*, 89–103.

Jouini, R., Houaidia, C., & Saidane, L. A. (2023). Hidden Markov model for early prediction of the elderly's dependency evolution in ambient assisted living. *Annals of Telecommunications, 78*, 599–615.

Laeven, L., & Valencia, F. (2018). Systemic banking crises revisited. IMF Working Paper 18/206. International Monetary Fund. IMF.

Levinson, S. E., Rabiner, L. R., & Sondhi, M. M. (1983). An introduction to the application of the theory of probabilistic functions of a Markov process to automatic speech recognition. *Bell System Technical Journal, 62*, 1035–1074.

Malhotra, R., Singla, C., & Farooque, D. (2022). Comparison of hidden Markov model with other machine learning techniques in software defect prediction. *Proceedings of 2022 IEEE 7th International Conference for Convergence in Technology (I2CT)* (pp. 1–5). Pune, India: IEEE.

McCullagh, P., & Nelder, J. A. (1989). *Generalized linear models* (2nd ed.). London: Chapman and Hall/CRC.

Mews, S., Koslik, J.-O., & Langrock, R. (2025). How to build your latent Markov model: The role of time and space. *Statistical Modelling, 25*, 481–507.

Mor, B., Garhwal, S., & Kumar, A. (2021). A systematic review of hidden Markov models and their applications. *Archives of Computational Methods in Engineering, 28*, 1429–1448.

Pigini, C. (2021). Penalized maximum likelihood estimation of logit-based early warning systems. *International Journal of Forecasting, 37*, 1156–1172.

R Core Team. (2025). *R: A Language and Environment for Statistical Computing*. Vienna, Austria: R Foundation for Statistical Computing.

Rothenbuehler, P., Runge, J., Garcin, F., & Faltings, B. (2015). Hidden Markov models for churn prediction. *Proceedings of 2015 SAI Intelligent Systems Conference (intellisys)* (pp. 723–730). London, UK: IEEE.

Saito, T., & Rehmsmeier, M. (2015). The precision-recall plot is more informative than the ROC plot when evaluating binary classifiers on imbalanced datasets. *PloS One, 10*, 1–21.

Schrodt, P. A. (1997). Early warning of conflict in southern Lebanon using hidden Markov models. In H. Starr (Ed.), *Proceedings of The Understanding and Management of Global Violence: New Approaches to Theory and Research on Protracted Conflict*, pp. 131–162. St. Martin's Press, New York.

Schwarz, G. (1978). Estimating the dimension of a model. *The Annals of Statistics, 6*, 461–464.

Tancini, D., Bartolucci, F., & Pandolfi, S. (2024). Bayesian hidden Markov models for early warning. *Proceedings of the 38th International Workshop on Statistical Modelling (IWSM)* pp. 55–61. Durham, UK: Durham University.

Umar, Z., Pradhan, B., Ahmad, A., Jebur, M. N., & Tehrany, M. S. (2014). Earthquake induced landslide susceptibility mapping using an integrated ensemble frequency ratio and logistic regression models in West Sumatera Province, Indonesia. *Catena, 118*, 124–135.

Valverde-Albacete, F. J., & Peláez-Moreno, C. (2014). 100% classification accuracy considered harmful: The normalized information transfer factor explains the accuracy paradox. *PloS One, 9*, 1–10.

Welch, L. R. (2003). Hidden Markov models and the Baum-Welch algorithm. *IEEE Information Theory Society Newsletter, 53*, 10–13.

Wooldridge, J. M. (2010). *Econometric analysis of cross section and panel data*. Cambridge, MA: MIT press.

Yip, C. F., Ng, W. L., & Yau, C. Y. (2018). A hidden Markov model for earthquake prediction. *Stochastic Environmental Research and Risk Assessment, 32*, 1415–1434.

Youden, W. J. (1950). Index for rating diagnostic tests. *Cancer, 3*, 32–35.

Zucchini, W., MacDonald, I. L., & Langrock, R. (2016). *Hidden Markov models for time series: An introduction using R* (2nd ed.). Boca Raton: Chapman and Hall/CRC.

A Regularized EWS for Banking Crises: A Grouped Fixed Effects Approach

Claudia Pigini and Alessandro Pionati

Abstract This chapter illustrates how the regularization introduced by the Grouped Fixed Effects (GFE) approach can be effectively applied to early warning identification and financial crisis forecasting. We use the well-established Macro History Database, which compiles historical data on financial turmoil over the past century. The GFE estimator proves effective in detecting early warnings, providing estimates comparable to those of the fixed effects method—the most common alternative in the literature—while mitigating the complete separation problem and relying on a more parsimonious specification. We further assess the predictive performance of the GFE approach and demonstrate its usefulness in forecasting first-ever crises. GFE successfully identifies such crises and shows good forecasting ability. Finally, we show that GFE outperforms popular machine learning algorithms in this context, while offering the added benefit of interpretability.

Keywords Early warning systems · Grouped fixed effects · Rare events

1 Introduction

The prediction of financial crises is a key objective for policy-makers. There is broad consensus in the literature that identifying early warning signals provides valuable guidance for preventing critical events (Borio & Lowe, 2002; Kaminsky & Reinhart, 1999). Early Warning Systems (EWS) are designed to generate timely and informative signals related to potential crises, thereby enabling policy-makers to implement appropriate responses in advance.

Supplementary Information The online version contains supplementary material available at https://doi.org/10.1007/978-3-032-24141-2_4.

C. Pigini (✉) · A. Pionati
Marche Polytechnic University, Ancona, Italy
e-mail: c.pigini@staff.univpm.it

A. Pionati
e-mail: a.pionati@staff.univpm.it

F. Bartolucci et al. (eds.), *Models for Longitudinal Data with Applications to Early Warning Systems*, Springer Proceedings in Mathematics & Statistics 553,
https://doi.org/10.1007/978-3-032-24141-2_4

Econometric identification of EWS has traditionally relied on parametric models like logit and probit (Alessi & Detken, 2018; Allaj & Sanfelici, 2023; Caggiano et al., 2016; Davis & Karim, 2008; Filippopoulou et al., 2020; Pigini, 2021), also with a hidden Markov specification for the unobserved heterogeneity (see Chap. Exploring Binary Regression and Hidden Markov Models for Early Warning Systems of this book). However, recent research has highlighted the superior predictive power of Machine Learning Algorithms (MLA) (Nazareth & Reddy, 2023). Common ML methods in the EWS literature include tree-based models (Alessi & Detken, 2018), neural networks (Tölö, 2020), support vector machines (Gogas et al., 2018), Lasso Logit (Lang et al., 2018), and AdaBoost (Casabianca et al., 2022).

While parametric models provide clear, interpretable estimates, MLA methods prioritize prediction over transparency. Among traditional parametric approaches, the one-way fixed effects model is particularly valued for addressing country-specific, time-invariant unobserved heterogeneity (UH) (Dawood et al., 2017; Pigini, 2021; Schularick & Taylor, 2012; Shen & Hsu, 2022).

In this chapter, we evaluate the performance of the recently proposed Grouped Fixed Effects (GFE) estimator (Bonhomme et al., 2022) in its dual capacity to identify early warning signals and predict upcoming financial crises. The GFE method adopts a data-driven strategy that uses k-means clustering to generate low-dimensional group-specific fixed effects—rather than country-specific ones—and combines this with standard parametric maximum likelihood estimation. This approach yields a more parsimonious model specification and effectively addresses issues of complete separation that often hinder prediction of first-ever events.

We present the following main findings. First, our results are consistent with the existing EWS literature, as we identify the same key indicators highlighted in prior studies. Notably, we reaffirm the role of rapid domestic credit growth as a leading early warning indicator, lending further empirical support to the theoretical frameworks proposed in Schularick and Taylor (2012), Jordà et al. (2017), Aikman et al. (2015), among others.

Second, we show that financial crises can be predicted using both parametric and machine learning models, albeit with different levels of performance. In particular, we find that the GFE estimator exhibits strong predictive accuracy and outperforms the fixed effects logit model in terms of the number of crises correctly predicted. Furthermore, we demonstrate that the GFE approach is particularly well suited for forecasting first-ever crises—a task where traditional fixed effects models fall short.

The code employed for data elaborations is implemented for the software R (R Core Team, 2025). Replication material for this chapter is available as supplementary files to the book. This chapter is structured as follows: Sect. 2 introduces the models used in the empirical analysis. Section 3 describes the Jordà-Schularick-Taylor Macro History Database. Section 4 presents the estimation results. Section 5 details the forecast exercise. Section 6 presents two robustness checks. Finally, Sect. 7 concludes.

2 Models and Estimators

In this section, we present the models and estimators employed in the analysis. We briefly discuss the standard one-way fixed effects (FE henceforth) approach for binary choice models and its limits in presence of an outcome variable with limited variability, as in the crises prediction framework. Then, we discuss the FE approach and explain the reasons why its use is advocated in the present context.

2.1 *Fixed Effects Approach*

For $i = 1, \ldots, N$ and $t = 1, \ldots, T$, let us define the following baseline model:

$$y_{it} = 1(\delta_0 y_{i,t-1} + x'_{i,t-1}\beta_0 + \alpha_{i0} + u_{it} > 0), \tag{1}$$

where y_{it} is a dummy variable equal to 1 when country i faces a crisis in time t, $1(\cdot)$ is the indicator function, and $y_{i,t-1}$ is the dependent variable lagged for a period; the associated parameter δ_0 is the state dependence parameter and gives the dimension of the association between the dependent variable and its lagged value. α_{i0} parameterizes individual-specific unobserved traits and $x_{i,t-1}$ denotes a set of k individual-specific covariates associated with a conformable vector of unknown parameters β_0. u_{it} is an independent identically distributed error component and y_{i0} is known. We assume that the error term follows a logistic distribution, meaning that model (1) is a standard dynamic logit model. In our applications, we also make use of a static logit model, which coincides with model 1 when $\delta_0 = 0$.

Estimation of model (1) is usually carried out with the FE approach: this procedure consists of estimating both the parameters of interest (δ_0, β_0) and the nuisance parameters referring to the UH (α_{i0}) using ML. Under minimal assumptions,[1] usually met in empirical applications, the FE estimator is consistent and asymptotically normally distributed. As it is well known in the related literature (Hahn & Kuersteiner, 2011), fixed effects estimation in binary choice models is subject to the incidental parameters problem (Lancaster, 2000; Neyman & Scott, 1948). If the number of time periods is fixed, the ML estimator of the regression parameters is not consistent, due to the estimation noise in the nuisance parameters. Instead, with $T \to \infty$ and N, T growing at the same rate (Li et al., 2003), the ML estimator is generally consistent, but exhibit an asymptotic bias: analytical (Hahn & Kuersteiner, 2011) and jackknife (Dhaene & Jochmans, 2015) bias corrections are available.

A significant challenge encountered by practitioners when dealing with binary choice panel data models, where limited transitions are observed in the outcome variable, pertains to the problem of complete separation (Albert & Anderson, 1984). For instance, consider a panel dataset describing rare events—such as the Jordà-

[1] We refer the reader to Hahn and Kuersteiner (2011) for details on the assumptions for the asymptotic properties of the FE estimator in the context of dynamic binary choice models.

Schularick-Taylor Macro History Database in exam—in which we observe severe sparsity of one of the two binary outcomes: countries are characterized by few transitions in the dependent variable and some of them do not experience transitions at all. For countries that do not experience any crises, FE estimation is critical, as the estimate of the individual fixed effect does not exist finite for the involved country: this happens because whenever data are not informative on the related fixed effect, the associated likelihood is monotone for that parameter. In these cases, ML estimates for the model parameters can be obtained at the cost of losing the portion of non-informative data, because in practice problematic countries are automatically removed from the dataset by the statistical software.

In our framework, the complete separation problem may acquire unexpected severity when trying to forecast first-ever crises. Let us define a first-ever event as the first appearance of a financial crisis for country i at time $T + 1$ for the first time. FE estimation of α_i using the observations $t = 1, \ldots, T$ leads to a non-finite estimate. In light of this, the predicted outcome at $T + 1$ will be systematically 0. If the number of countries that face financial distress only out of sample is large, forecasting performance of the FE estimator is bound to be poor. In our view, the GFE approach can introduce relevant innovations in this respect.

2.2 *Grouped Fixed Effects Approach*

The main idea behind the GFE approach is that a set of group-specific parameters can offer an approximation of time-invariant latent traits. With reference to model (1), the GFE approach estimates $\hat{\alpha}_k$ for $k = 1, \ldots, K$ grouped fixed effects, where $K \ll N$. Since the number of estimated UH parameters is smaller with respect to FE, we refer to the GFE as a *regularization* procedure (Pigini et al., 2025). The number of groups K is neither known nor fixed, and represents the number of approximating points for the continuous UH. In fact, $K \to \infty$ as $N \to \infty$, with $K/N \to 0$. This makes the GFE proposed by Bonhomme et al. profoundly different from impactful contributions in the literature on discrete structures for unobservable traits (Bonhomme & Manresa, 2015).

GFE estimation is computed by means of a simple two-step procedure: in the first step *k-means* clustering to find K groups of subjects is performed using the individual averages of the regressors.[2] In the second step, a set of K group-specific dummies is created, which then enters the model specification and grouped fixed effects are estimated via ML, together with other regression parameters. Since in practice the number of groups is unknown, Bonhomme et al. (2022) developed a rule in order to find the optimal K. The rule consists of an iterative procedure that sequentially

[2] Bonhomme et al. (2022) employ also the outcome variable in the clustering procedure, as in normal settings it entails information on unobservable traits. Following simulation results in Pigini et al. (2025), we conjecture that a sparse binary outcome variable may induce numerical noise without providing much information. We then perform the clustering step using explanatory variables only.

increases the number of groups until the variability of sample moments is of the same order as the objective function of the *k-means* problem. In *formulae*:

$$\hat{K} = \min_{K \geq 1}\{K : \hat{Q}(K) \leq \gamma_k \hat{V}_{h_i}\}, \tag{2}$$

where $Q(\cdot)$ is the objective function of the *kmeans* problem,[3] V_{h_i} is the variability of the moments h_i, and $\gamma_k \in (0, 1]$ is a user-specified parameter.

Bonhomme et al. (2022) study the asymptotic properties of the GFE estimator: provided that minimal assumptions are met,[4] it is consistent and, for N and T that grow at the same rate, is asymptotically biased and normally distributed. The bias comes from three sources: first, there is a bias component induced by the discretization of the UH through *kmeans*. Then, there is an incidental parameter bias caused by the limited number of individual moments employed in the clustering procedure. Finally, we have the aforementioned incidental parameter problem that affects non-linear models. Bonhomme et al. (2022) show that this bias is constant, provided the chosen number of groups is at least of the order of $\sqrt{T}$. Under this condition,the GFE estimator can be debiased using a jackknife procedure (Dhaene & Jochmans, 2015), if standard additional assumptions required for the applicability of this method are met.

The choice of γ_k in rule (2) is decisive because it indirectly controls the number of groups and, in turn, the goodness of approximation and the severity of the incidental parameters bias. Higher values of γ_k lead to a lower number of groups, worsening the discretization, but keeping the number of grouped fixed effects under control; *vice versa*, a small γ_k guarantees a good approximation at the cost of a larger incidental parameters bias. We discuss in Sect. 4 how to practically select the optimal γ_k in our study.

In our opinion, the proposed approach could be highly useful for the analysis of datasets where the outcome variable exhibits few state transitions or describes rare events. As discussed above, these settings are those where FE estimation is very likely to run into complete separation issues. Instead, the GFE approach intrinsically reduces the number of complete separation instances: the reduced number of fixed effects to estimate leads to a decrease in the likelihood of dealing with a set of individuals clustered in a group where no variability in the outcome variable is observed. As a consequence, the number of observations dropped due to complete separation is lower.

---

[3] In practice, $Q(\cdot)$ is the within sum of squares computed on the $\hat{K}$ groups.

[4] We remark here that, beyond standard assumptions required for a well-posed maximum likelihood problem, GFE estimator relies on two peculiar assumptions. The first one parameterizes the UH as a smooth, Lipschitz-continuous function of latent individual and time-specific traits. The second assumption imposes that moments used in the clustering procedure should be informative on latent traits of individuals and time occasions. In the limit, the difference between moments and the unspecified UH vanishes at a parametric rate. We refer the reader to Bonhomme et al. (2022) for all the details.

This feature turns out to be convenient when the goal of the analysis is to predict rare events. As discussed above, FE estimator meets substantial limits in prediction of first-ever crises. GFE has a strong advantage: since countries are clustered in groups, it happens that countries affected by at least one crisis and countries never affected by a crisis may mingle in the same cluster. If this happens, GFE produces a finite estimate of the group-specific intercept thus expressing a meaningful prediction for the crisis probability out of sample for countries that do not experiment financial crises at all.

The key distinction of the GFE approach lies in its ability to predict first-time crises, setting it apart from the traditional fixed effects approach. Additionally, the GFE method differs significantly from the machine learning techniques discussed in this paper, as it consists of a parametric approach where the complete interpretability of parameters of interest as early warnings is guaranteed.

3 Data

The dataset used in this analysis is the well-studied Jordà-Schularick-Taylor Macro History Database (Jordà et al., 2017; Schularick & Taylor, 2012) which collects annual observations on an extensive number of macroeconomic and financial variables in the years spanning from 1870 to 2020 for 18 countries.

The integrations made to the dataset are minimal to address negligible data gaps. In the final dataset (1954–2016), missing values are handled as follows: for Japan (1953–1956) and Norway (1966), where short-term interest rate data are missing (five observations), we use the risk-free rate from the Jordà-Schularick-Taylor Macro History Database as a proxy, following Bluwstein et al. (2023). Missing public debt data for Denmark (1952, 1957–1959, 1997) and Belgium (1980–1981) are retrieved from the International Monetary Fund database. Specifically, for Denmark, we use central government debt as a proxy for 1952 and 1957–1959, while for Denmark (1997) and Belgium (1980–1981), general government debt is used, consistent with the Jordà-Schularick-Taylor Macro History Database. For each country and each year, the dataset has a variable indicating whether a crisis happened in that particular year: the critical event has a sparse nature since the full dataset contains 88 crises over 2,718 observations. In order to define a critical event, Schularick and Taylor (2012) use the following criterion: "financial crises [are] events during which a country's banking sector experiences bank runs, sharp increases in default rate accompanied by large losses of capital that result in public intervention, bankruptcy, or forced merger of financial institutions." The definition of financial crisis by Schularick and Taylor (2012) roots in previous literature (Bordo et al., 2001; Reinhart & Rogoff, 2009) and has been widely accepted by economic historians in the related countries (Schularick & Taylor, 2012). In order to check for robustness of our results, we perform a sensitivity analysis using the definition of financial crisis outlined in Baron et al. (2021).

Due to its considerable length, the dataset under consideration contains some missing values, particularly in the observations from the late nineteenth and early twentieth centuries. We have chosen to focus our analysis on the latter half of the twentieth century, following World War II, for two main reasons: although the Jordà-Schularick-Taylor Macro History Database extends to the pre-WW2 era, the number of missing values for most of the variables employed in our analysis is substantial; by concentrating on more recent periods, we can rely on a balanced dataset free of missing observations. Secondly—in line with Bluwstein et al. (2023)—narrowing our focus to more recent events allows us to provide conclusions that are more relevant for contemporary policy-making.

We use two datasets of different lengths depending on whether the goal is estimation or prediction. For the estimation procedure, we employ data spanning from 1954 to 2016, resulting in a dataset with 1,134 observations. The analysis stops in 2016 due to the unavailability of data for some countries. When forecasting banking crises, we use observations from 1954 to 2008, as this is the last year in which Schularick and Taylor (2012) identify a critical event. Consequently, we cannot meaningfully extend the forecast horizon beyond this point, yielding a panel composed of a total of 990 observations.

We select the relevant explanatory variables by drawing on the existing literature on crisis forecasting (Bluwstein et al., 2023; Greenwood et al., 2022; Laeven & Valencia, 2018; Pigini, 2021; Schularick & Taylor, 2012) and based on data availability. The regressors used in our analysis include (see also Table 1 for a summary): investment, public debt, domestic credit, debt service ratio (calculated as credit $\times$ long-term interest rate over GDP), slope of the domestic yield curve (computed as the difference between short- and long-term interest rates), real consumption per capita, consumer price index (CPI), GDP per capita, and current account. Investment, public debt, and current account are expressed as ratios to GDP, while domestic credit, consumption, CPI, public debt over GDP, and GDP are considered in terms of growth rates. The domestic yield curve and the debt service ratio, on the other hand, are used in levels. These transformations are necessary to ensure comparability across variables and to address potential issues of non-stationarity. All explanatory variables are taken at time $t-1$, reflecting the reasonable assumption that banking crises at time t stem from macroeconomic imbalances in the previous year.

Credit growth plays a particularly significant role in our analysis, as it has been consistently highlighted in the literature as a strong predictor of critical episodes (Bluwstein et al., 2023; Greenwood et al., 2022; Schularick & Taylor, 2012). Its importance is emphasized in the context of a finance-driven explanation of economic crises: credit booms are often associated with excessive risk-taking and heightened financial instability (Aikman et al., 2015), which has led some scholars (Aliber et al., 2023) to describe financial crises as credit booms that have gone away.

The slope of the yield curve warrants specific attention: under normal conditions, this indicator is typically positive, indicating that long-term interest rates exceed short-term ones. This reflects the risk premium for holding long-term assets. However, an inverted yield curve may serve as a useful predictor of financial turmoil for two key reasons: first, when the curve turns negative, it suggests a premium for

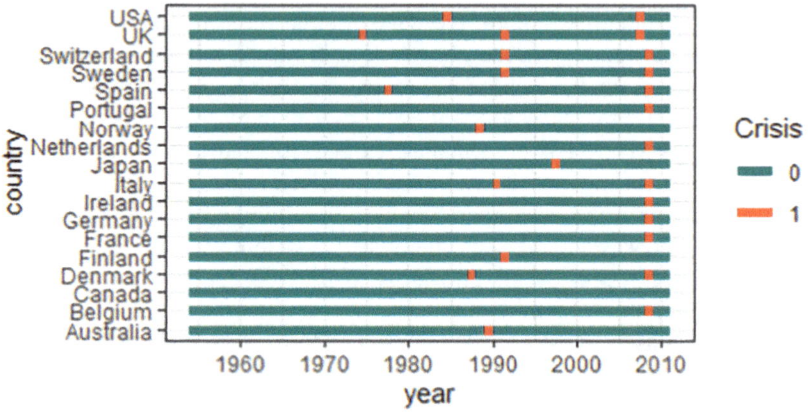

Fig. 1 Financial crises by year and country: red squares indicate critical events

short-term risk, indicating that investors are seeking out riskier opportunities (Taylor, 2009). Second, a flattened yield curve is linked to reduced net interest margins, which can weaken the banking sector and potentially constrain the supply of credit (Borio et al., 2017).

Within the selected time span, 25 financial crises are identified. Figure 1 provides a graphical overview of their frequency: with the exception of Canada, all countries in the sample experience at least one crisis, with a notable concentration in 2008, a year in which 11 critical events occur. Accordingly, the Great Crisis holds a central position in our analysis due to its broad and disruptive impact. An earlier, though shorter, wave of financial instability can be observed in the early 1990s, during which five countries faced episodes of economic distress.

The year 2007 may be viewed as a precursor to the subsequent financial turbulence: in that year, two major economies—United Kingdom and United States—were already hit by banking crises. Notably, both countries had previously experienced at least one episode of banking distress (the UK, two).

Six countries in the dataset—Belgium, Germany, France, Portugal, the Netherlands, and Ireland—encountered their first and only crisis in 2008. Their inclusion is particularly important within our framework, as first-time crises cannot be captured using the standard FE approach due to the complete separation issue. Their presence in the dataset is therefore essential to demonstrate the practical advantages of the GFE procedure in forecasting first-time financial crises.

4 Estimation Results

In this section, we present the estimation results. The model employed is a logit model, a standard choice in the literature for early warning signal identification (Demirgüç-Kunt & Detragiache, 1998, 2005). As a baseline, we adopt a *static* spec-

Table 1 Descriptive statistics

Name	Transformation	Min.	1st Qu.	Median	Mean	3rd Qu.	Max.	SD
Dependent variable								
Crisis	Levels	0	0	0	0.022	0	1	0.146
Explanatory variables								
Investment	GDP ratio	0.128	0.206	0.227	0.233	0.255	0.389	0.040
Public debt	Growth rate of GDP ratio	−0.383	−0.045	0.000	0.011	0.052	0.670	0.097
Domestic credit	Growth rate	−0.239	0.010	0.050	0.052	0.092	0.300	0.063
Debt service ratio	Levels	−0.000	0.000	0.002	0.080	0.016	3.513	0.284
Domestic yield curve slope	Levels	−10.925	0.254	1.181	1.146	1.987	14.150	1.846
Real Consumption per capita	Growth rate	−0.092	0.008	0.022	0.023	0.037	0.109	0.025
CPI	Growth rate	−0.046	0.017	0.030	0.042	0.057	0.321	0.041
GDP per capita	Growth rate	−0.086	0.016	0.030	0.031	0.047	0.279	0.028
Current account	GDP ratio	−0.132	−0.021	−0.001	0.001	0.020	0.160	0.037

Years in exam 1954–2016, 18 countries. Crisis refers to critical events as defined in Schularick and Taylor (2012). Growth rates computed on a yearly basis. Real GDP per capita is measured on PPP ification, while a dynamic version is introduced in Sect. 5 to produce forecasts. We estimate three alternative specifications: the pooled model, which completely ignores heterogeneity; the FE model, which fully captures country-specific heterogeneity; and the GFE model, estimated across a range of γ_k values, allowing for grouped effects. To address the bias inherent in the FE estimator, we apply the analytical bias correction proposed by Hahn and Kuersteiner (2011).

Although the GFE framework theoretically permits the application of jackknife bias correction, this option proves impractical due to the high sparsity of the dependent variable: subsampling the data frequently leads to subsamples where the outcome variable does not vary, resulting in numerical instability when computing the jackknife estimator as a linear combination of estimators computed within the multiple subsamples. For the same reason, the jackknife correction for the FE estimator is also omitted.

Estimation results are presented in Table 2, where we adopt a one-lag specification and report the estimated coefficients for the static logit model, together with the corresponding standard errors. The pooled model yields estimates that differ from those obtained using estimators accounting for heterogeneity. The analytical bias correction applied to the FE estimator has only a limited impact on the coefficients, indicating that the incidental parameters bias is relatively minor in this setting. Varying the value of γ_k in the GFE specification produces different results: for lower values of the hyperparameter, the estimates closely resemble those from the FE model. When $\gamma_k = 1$—its maximum value, which corresponds to the highest possible number of groups—the GFE procedure identifies eight groups, meaning that eight grouped fixed effects are included in the model. The "Dropped" row in Table 2 shows the number of countries excluded due to the complete separation problem. The FE and GFE estimators (for all γ_k except 0.5 and 1) must exclude one country to produce estimates, while the GFE with higher γ_k values does not require any exclusion.

All estimators agree in identifying the coefficients for public debt, the domestic yield curve, and consumption as statistically significant. However, only the pooled model and the GFE with $\gamma_k = 0.5$ fail to detect significance in credit growth and investment. The positive sign on the investment coefficient reflects the typical financial overheating that characterizes the run-up to crises, while the negative coefficient for consumption indicates that real and financial components of the economy tend to follow diverging paths as a crisis approaches. The growth rate of the debt-to-GDP ratio is associated with a negative and statistically significant coefficient. This suggests that a decline in the growth rate of the ratio is linked to an increased likelihood of a financial crisis. This result aligns with the explanation provided by Schularick and Taylor (2012): during periods of economic overheating, rising GDP causes the debt-to-GDP ratio to fall. At the same time, debt levels remain contained, as borrowing is relatively inexpensive due to low interest rates and strong economic activity. As a result, in the lead-up to a crisis, the ratio tends to decline. When the crisis eventually unfolds, the ratio rises again, primarily due to government intervention aimed at mitigating contagion effects through the banking sector. The coefficient for credit growth is positive, confirming the theory advanced by Schularick and Taylor

Table 2 Logit model for EWS: Model 1 with $\delta_0 = 0$

	Pooled	FE	BC	GFE					
γ_k				0.01	0.05	0.1	0.25	0.5	1
Investment (–1)	0.077	0.184	0.172	0.184	0.186	0.169	0.159	0.132	0.149
	(0.052)	(0.085)	(0.085)	(0.085)	(0.085)	(0.082)	(0.083)	(0.083)	(0.076)
Debt/GDP (–1)	–0.074	–0.068	–0.064	–0.068	–0.066	–0.067	–0.064	–0.066	–0.065
	(0.027)	(0.031)	(0.031)	(0.031)	(0.031)	(0.031)	(0.029)	(0.029)	(0.028)
Credit (–1)	0.050	0.075	0.079	0.075	0.075	0.074	0.075	0.062	0.069
	(0.033)	(0.037)	(0.037)	(0.037)	(0.037)	(0.037)	(0.036)	(0.035)	(0.036)
Debt service r. (–1)	0.688	1.116	1.116	1.116	1.145	1.147	1.085	1.104	1.002
	(0.65)	(1.11)	(1.08)	(1.11)	(1.11)	(1.09)	(1.04)	(1.07)	(1.08)
Domestic yield curve (–1)	–0.386	–0.424	–0.418	–0.424	–0.426	–0.428	–0.400	–0.406	–0.403
	(0.12)	(0.13)	(0.13)	(0.13)	(0.13)	(0.13)	(0.12)	(0.12)	(0.12)
Consumption (–1)	–0.273	–0.359	–0.353	–0.359	–0.355	–0.357	–0.355	–0.322	–0.339
	(0.121)	(0.133)	(0.134)	(0.133)	(0.133)	(0.134)	(0.132)	(0.133)	(0.134)
CPI (–1)	–0.097	–0.116	–0.111	–0.116	–0.116	–0.111	–0.108	–0.103	–0.106
	(0.068)	(0.072)	(0.071)	(0.072)	(0.072)	(0.072)	(0.072)	(0.072)	(0.072)
GDP (–1)	–0.002	0.049	0.044	0.049	0.048	0.048	0.049	0.038	0.049
	(0.111)	(0.105)	(0.109)	(0.105)	(0.105)	(0.106)	(0.106)	(0.112)	(0.112)
Current account/GDP (–1)	–0.096	–0.139	–0.140	–0.139	–0.135	–0.145	–0.141	–0.142	–0.153
	(0.061)	(0.089)	(0.089)	(0.089)	(0.089)	(0.089)	(0.087)	(0.087)	(0.083)
Dropped	0	1	1	1	1	1	1	0	0
K	0	0	0	18	17	16	14	11	8

"Pooled" is the pooled logit model; "FE" indicates the fixed effects estimator; "BC" is the analytical bias-corrected FE estimator following Hahn and Kuersteiner (2011); "GFE" reports the grouped fixed effects estimates for different values of γ_k. "Dropped" indicates the number of countries excluded due to the complete separation problem. K is the number of latent groups identified in the first step. $N = 18$, $T = 62$. Coefficients and standard errors for the following variables have been divided by 100: investment, public debt, credit, consumption, CPI, GDP growth, and current account to GDP

(2012). The coefficient on the slope of the domestic yield curve remains significant and negative, reinforcing the idea that an inverted yield curve is associated with a heightened probability of banking crises. The remaining coefficients are not statistically significant. To provide practical guidance, we recommend using $\gamma_k = 1$: this value of the hyperparameter yields estimates close to the bias-corrected FE model, avoids dropping countries from the sample, and results in a manageable number of fixed effects—eight in total.

5 Predicting Financial Crises

In this section, we assess the predictive performance of the models introduced above. To generate forecasts of financial crises, we divide the dataset into a training and a test set. The training set covers the period from 1954 to 2006. For the test set, we adopt a rolling window approach focused on the years 2007 and 2008, meaning that we report forecasting statistics for those 2 years. Although the Jordà-Schularick-Taylor Macro History Database includes crises prior to this period, we concentrate on the more recent systemic financial collapse due to its widespread and significant impact: 13 out of 18 countries in the dataset were affected, with 11 experiencing a crisis precisely in 2008.

Since we aim to predict a binary outcome, it is necessary to define a cut-off for the in-sample fitted probabilities. Specifically, for a given threshold c, the predicted outcome $\hat{p}_{it}$ for country i in year t is set to 1 if $\hat{p}_{it} > c$. To identify the optimal cut-off, we apply the following procedure: for a selected metric—such as the number of true positives (i.e., correctly identified crises)—we determine the value of c that maximizes this metric within the training set. The resulting optimal cut-off is then used to classify fitted probabilities in the test set. Naturally, choosing different metrics yields different optimal thresholds, and the most appropriate choice depends on the policy-maker's objective function. In the context of crisis forecasting, two types of errors may occur: issuing a crisis warning when none materializes (false alarm), or failing to predict an actual crisis (missed crisis). Both types of errors carry distinct and significant costs. Given the extreme sparsity of the outcome variable, selecting appropriate evaluation metrics is not straightforward. It is preferable to adopt statistics that account for the under-representation of crisis events. In this light, we present results based on three different metrics, each reflecting a distinct perspective on the policy-maker's prior beliefs. Specifically, we consider the following criteria for determining cut-offs: the number of in-sample true positives, the in-sample sum of specificity and sensitivity,[5] and the in-sample F1-score.[6]

[5] Specificity is defined as the ratio of true negatives to the sum of true negatives and false positives; sensitivity is the ratio of true positives to the sum of true positives and false negatives.

[6] The F1-score is calculated as $\frac{2 \times TP}{2 \times TP + FP + FN}$, where TP, FP, and FN denote true positives, false positives, and false negatives, respectively.

Tables 3, 4, 5, and 6 present out-of-sample forecast statistics for the described estimators across four specifications: the baseline static logit model used in Sect. 4, the same model augmented with predictors lagged at $t-2$, Model (1), and Model (1) with predictors lagged at $t-2$. Each table reports results based on cut-offs selected according to the three metrics outlined above. For each estimator, we track the four fundamental confusion matrix statistics (true positives, true negatives, false positives, and false negatives), the out-of-sample F1-score, the number of groups identified by GFE in the initial estimation step, and the number of countries dropped due to the complete separation problem.

The results across the four specifications are overall quite similar. In 2007, the United Kingdom and the United States experienced financial crises, so there are two positive events to predict. When the cut-off is chosen based on true positives or the sum of specificity and sensitivity, all estimators across all specifications successfully identify these two critical events. Using the sum of specificity and sensitivity leads to a more balanced performance overall, as the number of correctly predicted non-crisis events increases for all estimators, with GFE at $\gamma_k = 1$ performing best in terms of the F1-score. When the selected cut-off is the F1-score, forecasting performance generally declines, as no model manages to identify both crises: the pooled model fails to predict any crises, while in the one-lag specification only the FE and GFE estimators—regardless of γ_k—show relatively better performance.

The results for 2008 are markedly different: first, the crises are widespread, with 11 occurrences. Second, six countries experience their first-ever crises. The presence of first-ever crises severely hinders the forecasting ability of the FE estimator, since these six countries are considered non-critical by construction due to the complete separation problem. This means that FE fails to correctly predict over half of the crises. The GFE estimator holds a significant advantage because, as explained in Sect. 2, by regularizing the number of clusters, it is more likely that countries facing first-ever crises are grouped with regular countries, effectively eliminating the complete separation problem. Examining the 2008 columns in Tables 4 and 5, it is clear that when the cut-off is based on true positives, the pooled model and GFE with $\gamma_k = 1$ perform best, followed closely by GFE with other γ_k values. In this context, GFE with $\gamma_k = 1$ successfully predicts first-ever crises: for example, as shown in Tables 4 and 6, it correctly identifies four out of six first-ever crises (Belgium, France, Ireland, and the Netherlands), while misclassifying Germany and Portugal.[7]

In 2008, selecting cut-offs other than the number of true positives considerably reduces the predictive performance of all estimators. When the cut-off is the sum of specificity and sensitivity, GFE with $\gamma_k = 0.25$ achieves the best overall results, outperforming the pooled model across all specifications. Unfortunately, when the in-sample F1-score is used as the cut-off, the results are unsatisfactory for all estimators, and using this metric in similar contexts is not recommended.

The GFE estimator emerges as a superior tool for policy-makers compared to the FE and pooled logit models. While the quantification of EWS is similar across all

[7] Germany is wrongly classified as non-critical in 2008, and Portugal is alone in group 8, making its group fixed effect inestimable.

Table 3 Banking crises forecasts: Model (1) with $\delta_0 = 0$

	2007							2008						
	TP	TN	FP	FN	K	Dropped	F1	TP	TN	FP	FN	K	Dropped	F1
Cut-off: TP														
Pooled	2	1	15	0	–	0	0.211	9	1	6	2	–	0	0.692
FE	2	11	5	0	–	7	0.444	3	3	4	8	–	7	0.333
GFE 0.01	2	11	5	0	18	7	0.444	3	3	4	8	18	7	0.333
GFE 0.05	2	10	6	0	16	6	0.400	4	3	4	7	16	6	0.421
GFE 0.1	2	10	6	0	15	6	0.400	4	2	5	7	15	5	0.400
GFE 0.25	2	7	9	0	13	4	0.308	6	1	6	5	13	4	0.522
GFE 0.5	2	8	8	0	10	4	0.333	6	1	6	5	11	4	0.522
GFE 1	2	6	10	0	8	1	0.286	7	1	6	4	8	1	0.583
Cut-off: Sens + Spec														
Pooled	2	8	8	0	–	0	0.333	3	3	4	8	–	0	0.333
FE	2	13	3	0	–	7	0.571	2	4	3	9	–	7	0.250
GFE 0.01	2	13	3	0	18	7	0.571	2	4	3	9	18	7	0.250
GFE 0.05	2	13	3	0	16	6	0.571	2	4	3	9	16	6	0.250
GFE 0.1	2	13	3	0	15	6	0.571	3	4	3	8	15	5	0.353
GFE 0.25	2	13	3	0	13	4	0.571	3	4	3	8	13	4	0.353
GFE 0.5	2	9	7	0	10	4	0.364	3	4	3	8	11	4	0.353
GFE 1	2	14	2	0	8	1	0.667	1	4	3	10	8	1	0.133
Cut-off: F1-score														
Pooled	0	15	1	2	–	0	–	0	5	2	11		0	–
FE	1	16	0	1	–	7	0.667	0	5	2	11	–	7	–
GFE 0.01	1	16	0	1	18	7	0.667	0	5	2	11	18	7	–
GFE 0.05	1	15	1	1	16	6	0.500	0	5	2	11	16	6	–
GFE 0.1	1	15	1	1	15	6	0.500	0	6	1	11	15	5	–
GFE 0.25	0	16	0	2	13	4	–	0	7	0	11	13	4	–
GFE 0.5	1	16	0	1	10	4	0.667	0	6	1	11	11	4	–
GFE 1	1	16	0	1	8	1	0.667	0	6	1	11	8	1	–

"TP" true positives, "TN" true negatives, "FP" false positives, "FN" false negatives. K reports the number of groups for individuals obtained in the first step for GFE . "Dropped" is the number of countries removed due to complete separation. "F1" is the out-of-sample F1-score. Estimators: "Pooled" is the pooled model, "FE" is standard FE estimator, "GFE" is standard GFE with different values of γ_k parameter. "–" denotes unavailable statistics. "Cut-off" indicates the metric optimized on the training set to get the cut-off for fitted probabilities

Table 4 Banking crises forecasts: Model (1) with $\delta_0 = 0$ and $x_{i,t-2}$

	2007							2008						
	TP	TN	FP	FN	K	Dropped	F1	TP	TN	FP	FN	K	Dropped	F1
Cut-off: TP														
Pooled	2	3	13	0	–	0	0.235	8	1	6	3	–	0	0.640
FE	2	11	5	0	–	7	0.444	3	3	4	8	–	7	0.333
GFE 0.01	2	11	5	0	18	7	0.444	3	3	4	8	18	7	0.333
GFE 0.05	2	11	5	0	17	7	0.444	4	3	4	7	16	6	0.421
GFE 0.1	2	10	6	0	16	6	0.400	4	3	4	7	15	6	0.421
GFE 0.25	2	7	9	0	13	4	0.308	6	1	6	5	13	4	0.522
GFE 0.5	2	7	9	0	11	4	0.308	6	1	6	5	11	4	0.522
GFE 1	2	6	10	0	8	1	0.286	8	1	6	3	8	1	0.640
Cut-off: Sens + Spec														
Pooled	2	6	10	0	–	0	0.286	3	3	4	8	–	0	0.333
FE	2	13	3	0	–	7	0.571	2	4	3	9	–	7	0.250
GFE 0.01	2	13	3	0	18	7	0.571	2	4	3	9	18	7	0.250
GFE 0.05	2	13	3	0	17	7	0.571	2	4	3	9	16	6	0.250
GFE 0.1	2	13	3	0	16	6	0.571	2	4	3	9	15	6	0.250
GFE 0.25	2	11	5	0	13	4	0.444	4	4	3	7	13	4	0.444
GFE 0.5	2	10	6	0	11	4	0.400	4	3	4	7	11	4	0.421
GFE 1	2	14	2	0	8	1	0.667	1	4	3	10	8	1	0.133
Cut-off: F1-score														
Pooled	0	16	0	2	–	0	–	1	5	2	10	–	0	0.143
FE	1	14	2	1	–	7	0.400	0	5	2	11	–	7	–
GFE 0.01	1	14	2	1	18	7	0.400	0	5	2	11	18	7	–
GFE 0.05	1	14	2	1	17	7	0.400	0	5	2	11	16	6	–
GFE 0.1	1	14	2	1	16	6	0.400	0	4	3	11	15	6	–
GFE 0.25	0	15	1	2	13	4	–	0	5	2	11	13	4	–
GFE 0.5	0	15	1	2	11	4	–	0	5	2	11	11	4	–
GFE 1	1	15	1	1	8	1	0.500	0	6	1	11	8	1	–

"TP" true positives, "TN" true negatives, "FP" false positives, "FN" false negatives. K reports the number of groups for individuals obtained in the first step for GFE . "Dropped" is the number of countries removed due to complete separation. "F1" is the out-of-sample F1-score. Estimators: "Pooled" is the pooled model, "FE" is standard FE estimator, "GFE" is standard GFE with different values of γ_k parameter. "–" denotes unavailable statistics. "Cut-off" indicates the metric optimized on the training set to get the cut-off for fitted probabilities

Table 5 Banking crises forecasts: Model 1

	2007							2008						
	TP	TN	FP	FN	K	Dropped	F1	TP	TN	FP	FN	K	Dropped	F1
Cut-off: TP														
Pooled	2	1	15	0	–	0	0.211	11	3	4	0	–	0	0.846
FE	2	11	5	0	–	7	0.444	3	5	2	8	–	7	0.375
GFE 0.01	2	11	5	0	18	7	0.444	3	5	2	8	18	7	0.375
GFE 0.05	2	10	6	0	16	6	0.400	4	5	2	7	16	6	0.471
GFE 0.1	2	9	7	0	15	6	0.364	5	3	4	6	15	5	0.500
GFE 0.25	2	5	11	0	13	4	0.267	6	3	4	5	13	4	0.571
GFE 0.5	2	8	8	0	10	4	0.333	6	3	4	5	11	4	0.571
GFE 1	2	6	10	0	8	1	0.286	7	3	4	4	8	1	0.636
Cut-off: Sens + Spec														
Pooled	2	7	9	0	–	0	0.308	3	6	1	8	–	0	0.400
FE	2	13	3	0	–	7	0.571	2	6	1	9	–	7	0.286
GFE 0.01	2	13	3	0	18	7	0.571	2	6	1	9	18	7	0.286
GFE 0.05	2	13	3	0	16	6	0.571	2	6	1	9	16	6	0.286
GFE 0.1	2	14	2	0	15	6	0.667	2	6	1	9	15	5	0.286
GFE 0.25	2	13	3	0	13	4	0.571	3	6	1	8	13	4	0.400
GFE 0.5	2	9	7	0	10	4	0.364	3	6	1	8	11	4	0.400
GFE 1	2	14	2	0	8	1	0.667	1	6	1	10	8	1	0.154
Cut-off: F1-score														
Pooled	0	15	1	2	–	0	–	0	6	1	11	–	0	–
FE	1	16	0	1	–	7	0.667	0	6	1	11	–	7	–
GFE 0.01	1	16	0	1	18	7	0.667	0	6	1	11	18	7	–
GFE 0.05	1	16	0	1	16	6	0.667	0	6	1	11	16	6	–
GFE 0.1	1	16	0	1	15	6	0.667	0	6	1	11	15	5	–
GFE 0.25	0	16	0	2	13	4		0	7	0	11	13	4	–
GFE 0.5	1	16	0	1	10	4	0.667	0	7	0	11	11	4	–
GFE 1	1	16	0	1	8	1	0.667	0	7	0	11	8	1	–

"TP" true positives, "TN" true negatives, "FP" false positives, " FN" false negatives. K reports the number of groups for individuals obtained in the first step for GFE . "Dropped" is the number of countries removed due to complete separation. "F1" is the out-of-sample F1-score. Estimators: "Pooled" is the pooled model, "FE" is standard FE estimator, "GFE" is standard GFE with different values of γ_k parameter. "–" denotes unavailable statistics. "Cut-off" indicates the metric optimized on the training set to get the cut-off for fitted probabilities

Table 6 Banking crises forecasts: Model (1) with $x_{i,t-2}$

	2007							2008						
	TP	TN	FP	FN	K	Dropped	F1	TP	TN	FP	FN	K	Dropped	F1
Cut-off: TP														
Pooled	2	3	13	0	–	0	0.235	10	3	4	1	–	0	0.800
FE	2	11	5	0	–	7	0.444	3	5	2	8	–	7	0.375
GFE 0.01	2	11	5	0	18	7	0.444	3	5	2	8	18	7	0.375
GFE 0.05	2	11	5	0	17	7	0.444	4	5	2	7	16	6	0.471
GFE 0.1	2	10	6	0	16	6	0.400	4	5	2	7	15	6	0.471
GFE 0.25	2	7	9	0	13	4	0.308	6	3	4	5	13	4	0.571
GFE 0.5	2	7	9	0	11	4	0.308	6	3	4	5	11	4	0.571
GFE 1	2	7	9	0	8	1	0.308	8	3	4	3	8	1	0.696
Cut-off: Sens + Spec														
Pooled	2	7	9	0	–	0	0.308	4	5	2	7	–	0	0.471
FE	2	13	3	0	–	7	0.571	2	6	1	9	–	7	0.286
GFE 0.01	2	13	3	0	18	7	0.571	2	6	1	9	18	7	0.286
GFE 0.05	2	13	3	0	17	7	0.571	2	6	1	9	16	6	0.286
GFE 0.1	2	13	3	0	16	6	0.571	4	6	1	7	15	6	0.500
GFE 0.25	2	10	6	0	13	4	0.400	5	6	1	6	13	4	0.588
GFE 0.5	2	9	7	0	11	4	0.364	2	6	1	9	11	4	0.286
GFE 1	2	14	2	0	8	1	0.667	1	6	1	10	8	1	0.154
Cut-off: F1-score														
Pooled	0	15	1	2	–	0	- -	0	6	1	11	–	0	–
FE	1	14	2	1	–	7	0.400	0	6	1	11	–	7	–
GFE 0.01	1	14	2	1	18	7	0.400	0	6	1	11	18	7	–
GFE 0.05	1	14	2	1	17	7	0.400	0	6	1	11	16	6	–
GFE 0.1	0	15	1	2	16	6	–	0	6	1	11	15	6	–
GFE 0.25	0	15	1	2	13	4	–	0	7	0	11	13	4	–
GFE 0.5	0	15	1	2	11	4	–	0	7	0	11	11	4	–
GFE 1	0	15	1	2	8	1	–	0	7	0	11	8	1	–

"TP" true positives, "TN" true negatives, "FP" false positives, "FN" false negatives. K reports the number of groups for individuals obtained in the first step for GFE . "Dropped" is the number of countries removed due to complete separation. "F1" is the out-of-sample F1-score. Estimators: "Pooled" is the pooled model, "FE" is standard FE estimator, "GFE" is standard GFE with different values of γ_k parameter. "–" denotes unavailable statistics. "Cut-off" indicates the metric optimized on the training set to get the cut-off for fitted probabilities

three approaches, the GFE method is more parsimonious than FE, as it estimates fewer fixed effects. Given that the GFE approach provides equal or better predictive performance than both pooled and FE models, it proves to be a more attractive option for practitioners focused on forecasting. Additionally, its ease of use and reliable estimates further enhance its appeal. Notably, the GFE method also allows for the forecasting of first-ever crises, a capability that FE models lack, making it an appealing tool for policy-makers.

6 Extensions

6.1 Predicting Financial Crises: Alternative Crisis Definition of Baron et al. (2021)

In this section, we conduct a robustness check by predicting financial crises as defined in Baron et al. (2021). The authors adopt a broader definition of crisis compared to that of Schularick & Taylor (2012): in their own words, "[a crisis is defined as] an episode in which the aggregate banking sector ability to intermediate funds is severely impaired." The crisis chronology provided by Baron et al. (2021) is based on data concerning bank equity prices and narrative accounts of bank failures and banking panics.[8] Their study covers 46 countries over the period 1870–2016. In parallel with what we do in Sect. 5, we focus on predicting crises within a specific rolling window. Our attention is on the more recent systemic financial collapse and its aftermath. Baron et al. (2021) identify the final crisis in their dataset in 2016, which enables us to extend the forecasting horizon: notably, they report at least one critical event occurring in 2007, 2008, 2010, 2011, and 2016.

Figure 2 provides a graphical representation of the chronology of financial turmoils. We use data spanning from 1964 to 2016; however, due to missing values in the outcome variable, we exclude four countries from the dataset (Italy, Japan, Finland, and Ireland). The set of explanatory variables remains the same as in Sect. 5. In the resulting dataset, we observe 30 crises over a 53-year period, 15 of which fall within the forecasting horizon. Under the framework of Baron et al. (2021), this results in a higher number of critical events compared to the chronology proposed by Schularick and Taylor (2012). In addition, we identify three first-ever crisis episodes, all occurring in 2008 (Spain, Denmark, and Belgium).

Tables 7, 8, 9, and 10 present forecast results over a 5-year horizon. We employ the same estimators and the same metrics for selecting the optimal cut-off as discussed in Sect. 5. The results of this robustness check broadly support the conclusions from the main analysis. The initial manifestation of the crisis in 2007, marked by financial

[8] Specifically, Baron et al. (2021) distinguish between two types of crises: bank equity crises and banking panic crises. The former are identified by a sharp annual drop in bank stock prices, exceeding 30%. The latter are determined through narrative evidence of severe bank runs. In our analysis, we use both types, combined into a single generic crisis indicator.

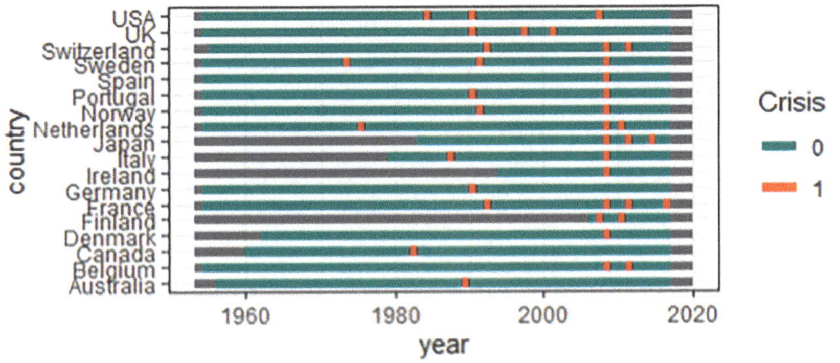

Fig. 2 Financial crises by year and country using the BVX definition: red squares indicate critical events; gray lines indicate missing values

distress in the US, is accurately predicted by all models when using either the true positives or the sum of specificity and sensitivity as cut-off criteria. In 2008, when nine crises erupt, the GFE model shows relatively balanced performance when the sum of specificity and sensitivity is used as the cut-off. Using the true positives rate instead leads to a higher number of correctly predicted crises, but at the cost of raising false alarms in non-crisis countries. These outcomes are consistent with the findings of Sect. 5. For what concerns the aftermath of the crisis, we highlight that all estimators deliver unsatisfactory predictions: while it is always possible to predict critical events by selecting the true positives as a cut-off, it is hard to achieve a balanced performance employing different metrics.

6.2 Comparison with Machine Learning Algorithms

In this section, we assess the predictive performance of MLAs in classifying financial crises using the Macro History Database. As discussed in Sect. 1, learning-based methods are increasingly becoming standard tools in the analysis of EWSs, and in some cases have demonstrated superior forecasting accuracy. However, their limited interpretability makes our proposed approach particularly attractive when predictive accuracy is comparable.

We tested five MLAs: random forest, decision tree, support vector machine, K-nearest neighbors, and logistic ridge regression. We employ the same methodology described above for the splitting of the dataset into training and test set and the choice of the optimal threshold. Moreover, we make use of the cross-validation procedure in order to select the optimal value of different hyper-parameters for the algorithms. Overall, MLAs do not display superior predictive performance on the dataset under examination compared to parametric models. While all MLAs tend to perform poorly when the cut-off is optimized based on the in-sample F1-score,

Table 7 Banking crises forecasts: Model (1) with $\delta_0 = 0$, definition of Baron et al. (2021)

	2007							2008							2010							2011							2016						
	TP	TN	FP	FN	K	CS	F1	TP	TN	FP	FN	K	CS	F1	TP	TN	FP	FN	K	CS	F1	TP	TN	FP	FN	K	CS	F1	TP	TN	FP	FN	K	CS	F1
Cut-off: TP																																			
Pool	1	1	12	0		0	0.14	8	0	5	1		0	0.73	1	0	13	0		0	0.13	3	1	10	0		0	0.38	1	0	13	0		0	0.13
ML	1	3	10	0		3	0.17	6	0	5	3		3	0.60	1	6	7	0		0	0.22	3	1	10	0		0	0.38	1	1	12	0		0	0.14
GFE 0.01	1	3	10	0	14	3	0.17	6	0	5	3	14	3	0.60	1	6	7	0	14	0	0.22	3	1	10	0	14	0	0.38	1	1	12	0	14	0	0.14
GFE 0.05	1	3	10	0	14	3	0.17	6	0	5	3	14	3	0.60	1	6	7	0	14	0	0.22	3	1	10	0	14	0	0.38	1	1	12	0	13	0	0.14
GFE 0.1	1	2	11	0	13	2	0.15	6	0	5	3	13	3	0.60	1	6	7	0	13	0	0.22	3	1	10	0	13	0	0.38	1	1	12	0	12	0	0.14
GFE 0.25	1	2	11	0	11	2	0.15	7	0	5	2	11	2	0.67	1	5	8	0	11	0	0.20	3	1	10	0	11	0	0.38	1	1	12	0	10	0	0.14
GFE 0.5	1	1	12	0	9	1	0.14	8	0	5	1	9	1	0.73	1	5	8	0	9	0	0.20	3	1	10	0	9	0	0.38	1	1	12	0	9	0	0.14
GFE 1	1	1	12	0	7	1	0.14	8	0	5	1	7	1	0.73	1	3	10	0	7	0	0.17	3	1	10	0	7	0	0.38	1	0	13	0	7	0	0.13
Cut-off: Sens + Spec																																			
Pool	1	5	8	0		0	0.20	3	0	5	6		0	0.35	0	11	2	1		0	–	0	10	1	3		0	–	0	8	5	1		0	–
ML	1	8	5	0		3	0.29	3	1	4	6		3	0.38	0	10	3	1		0	–	0	9	2	3		0	–	1	6	7	0		0	0.22
GFE 0.01	1	8	5	0	14	3	0.29	3	1	4	6	14	3	0.38	0	10	3	1	14	0	–	0	9	2	3	14	0	–	1	6	7	0	14	0	0.22
GFE 0.05	1	8	5	0	14	3	0.29	3	1	4	6	14	3	0.38	0	10	3	1	14	0	–	0	9	2	3	14	0	–	1	6	7	0	13	0	0.22
GFE 0.1	1	7	6	0	13	2	0.25	3	0	5	6	13	3	0.35	0	10	3	1	13	0	–	0	8	3	3	13	0	–	1	6	7	0	12	0	0.22
GFE 0.25	0	7	6	1	11	2	–	4	0	5	5	11	2	0.44	0	10	3	1	11	0	–	0	8	3	3	11	0	–	0	9	4	1	10	0	–
GFE 0.5	0	6	7	1	9	1	–	3	0	5	6	9	1	0.35	0	11	2	1	9	0	–	0	8	3	3	9	0	–	0	7	6	1	9	0	–
GFE 1	1	5	8	0	7	1	0.20	4	0	5	5	7	1	0.44	0	11	2	1	7	0	–	0	10	1	3	7	0	–	0	7	6	1	7	0	–
Cut-off: F1																																			
Pool	0	11	2	1		0	–	1	2	3	8		0	0.15	0	12	1	1		0	–	0	11	0	3		0	–	0	13	0	1		0	–
ML	0	13	0	1		3	–	1	4	1	8		3	0.18	0	12	1	1		0	–	0	10	1	3		0	–	0	13	0	1		0	–
GFE 0.01	0	13	0	1	14	3	–	1	4	1	8	14	3	0.18	0	12	1	1	14	0	–	0	10	1	3	14	0	-	0	13	0	1	14	0	–
GFE 0.05	0	13	0	1	14	3	–	1	4	1	8	14	3	0.18	0	12	1	1	14	0	–	0	10	1	3	14	0	-	0	13	0	1	13	0	–
GFE 0.1	0	9	4	1	13	2	–	1	3	2	8	13	3	0.17	0	12	1	1	13	0	–	0	10	1	3	13	0	–	0	13	0	1	12	0	-
GFE 0.25	0	13	0	1	11	2	–	1	3	2	8	11	2	0.17	0	12	1	1	11	0	–	0	10	1	3	11	0	–	0	13	0	1	10	0	–
GFE 0.5	0	13	0	1	9	1	–	0	5	0	9	9	1	–	0	13	0	1	9	0	–	0	11	0	3	9	0	–	0	13	0	1	9	0	–
GFE 1	0	13	0	1	7	1	–	0	5	0	9	7	1	–	0	13	0	1	7	0	–	0	11	0	3	7	0	–	0	11	2	1	7	0	–

"TP" true positives, "TN" true negatives, "FP" false positives, "FN" false negatives. K reports the number of groups for individuals obtained in the first step for GFE . "CS" is the number of countries removed due to complete separation. "F1" is the out-of-sample F1-score. Estimators: "Pooled" is the pooled model, "FE" is standard FE estimator, "GFE" is standard GFE with different values of γ_k parameter. "Cut-off" indicates the metric optimized on the training set to get the cut-off for fitted probabilities

Table 8 Banking crises forecasts: Model (1) with $\delta_0 = 0$ and $x_{i,t-2}$, definition of Baron et al. (2021)

	2007							2008							2010							2011							2016						
	TP	TN	FP	FN	K	CS	F1	TP	TN	FP	FN	K	CS	F1	TP	TN	FP	FN	K	CS	F1	TP	TN	FP	FN	K	CS	F1	TP	TN	FP	FN	K	CS	F1
Cut-off: TP																																			
Pool	1	0	13	0		0	0.13	9	0	5	0		0	0.78	1	3	10	0		0	0.17	3	0	11	0		0	0.35	1	0	13	0		0	0.13
ML	1	3	10	0		3	0.17	6	0	5	3		3	0.60	1	3	10	0		0	0.17	3	0	11	0		0	0.35	1	0	13	0		0	0.13
GFE 0.01	1	3	10	0	14	3	0.17	6	0	5	3	14	3	0.60	1	3	10	0	14	0	0.17	3	0	11	0	14	0	0.35	1	0	13	0	14	0	0.13
GFE 0.05	1	3	10	0	14	3	0.17	6	0	5	3	14	3	0.60	1	3	10	0	14	0	0.17	3	0	11	0	14	0	0.35	1	0	13	0	13	0	0.13
GFE 0.1	1	3	10	0	13	3	0.17	7	0	5	2	13	2	0.67	1	3	10	0	13	0	0.17	3	0	11	0	13	0	0.35	1	0	13	0	12	0	0.13
GFE 0.25	1	2	11	0	11	2	0.15	7	0	5	2	11	2	0.67	1	3	10	0	11	0	0.17	3	0	11	0	11	0	0.35	1	0	13	0	11	0	0.13
GFE 0.5	1	1	12	0	9	1	0.14	8	0	5	1	9	1	0.73	1	3	10	0	9	0	0.17	3	0	11	0	9	0	0.35	1	0	13	0	9	0	0.13
GFE 1	1	1	12	0	7	1	0.14	8	0	5	1	7	1	0.73	1	5	8	0	7	0	0.20	3	0	11	0	7	0	0.35	1	0	13	0	7	0	0.13
Cut-off: Sens + Spec																																			
Pool	1	4	9	0		0	0.18	3	1	4	6		0	0.38	0	11	2	1		0	–	0	9	2	3		0	–	0	11	2	1		0	–
ML	1	3	10	0		3	0.17	3	1	4	6		3	0.38	0	10	3	1		0	–	0	9	2	3		0	–	0	9	4	1		0	–
GFE 0.01	1	3	10	0	14	3	0.17	3	1	4	6	14	3	0.38	0	10	3	1	14	0	–	0	9	2	3	14	0	–	0	9	4	1	14	0	–
GFE 0.05	1	3	10	0	14	3	0.17	3	1	4	6	14	3	0.38	0	10	3	1	14	0	–	0	9	2	3	14	0	–	0	9	4	1	13	0	–
GFE 0.1	1	3	10	0	13	3	0.17	4	1	4	5	13	2	0.47	0	10	3	1	13	0	–	0	10	1	3	13	0	–	0	9	4	1	12	0	–
GFE 0.25	1	3	10	0	11	2	0.17	4	1	4	5	11	2	0.47	0	10	3	1	11	0	–	0	9	2	3	11	0	–	0	9	4	1	11	0	–
GFE 0.5	1	2	11	0	9	1	0.15	2	2	3	7	9	1	0.29	0	11	2	1	9	0	–	0	8	3	3	9	0	–	0	10	3	1	9	0	–
GFE 1	1	2	11	0	7	1	0.15	3	2	3	6	7	1	0.40	0	11	2	1	7	0	–	0	10	1	3	7	0	–	0	9	4	1	7	0	–
Cut-off: F1																																			
Pool	0	13	0	1		0	–	0	5	0	9		0	-	0	12	1	1		0	–	0	10	1	3		0	–	0	11	2	1		0	–
ML	0	13	0	1		3	–	2	5	0	7		3	0.36	0	12	1	1		0	–	0	10	1	3		0	–	0	12	1	1		0	–
GFE 0.01	0	13	0	1	14	3	–	2	5	0	7	14	3	0.36	0	12	1	1	14	0	–	0	10	1	3	14	0	–	0	12	1	1	14	0	–
GFE 0.05	0	13	0	1	14	3	–	2	5	0	7	14	3	0.36	0	12	1	1	14	0	–	0	10	1	3	14	0	–	0	11	2	1	13	0	–
GFE 0.1	0	12	1	1	13	3	–	2	5	0	7	13	2	0.36	0	12	1	1	13	0	–	0	10	1	3	13	0	–	0	11	2	1	12	0	–
GFE 0.25	0	12	1	1	11	2	–	2	5	0	7	11	2	0.36	0	12	1	1	11	0	–	0	10	1	3	11	0	–	0	11	2	1	11	0	–
GFE 0.5	0	13	0	1	9	1	–	1	5	0	8	9	1	0.20	0	12	1	1	9	0	–	0	10	1	3	9	0	–	0	11	2	1	9	0	–
GFE 1	0	13	0	1	7	1	–	0	5	0	9	7	1	–	0	12	1	1	7	0	–	0	10	1	3	7	0	–	0	11	2	1	7	0	–

"TP" true positives, "TN" true negatives, "FP" false positives, "FN" false negatives. K reports the number of groups for individuals obtained in the first step for GFE . "CS" is the number of countries removed due to complete separation. "F1" is the out-of-sample F1-score. Estimators: "Pooled" is the pooled model, "FE" is standard FE estimator, "GFE" is standard GFE with different values of γ_k parameter. "Cut-off" indicates the metric optimized on the training set to get the cut-off for fitted probabilities

Table 9 Banking crises forecasts: Model (1), definition of Baron et al. (2021)

	2007							2008							2010							2011							2016						
	TP	TN	FP	FN	*K*	CS	F1	TP	TN	FP	FN	*K*	CS	F1	TP	TN	FP	FN	*K*	CS	F1	TP	TN	FP	FN	*K*	CS	F1	TP	TN	FP	FN	*K*	CS	F1
Cut-off: TP																																			
Pool	1	1	12	0		0	0.14	8	1	4	1		0	0.76	1	0	13	0		0	0.13	3	1	10	0		0	0.38	1	0	13	0		0	0.13
ML	1	3	10	0		3	0.17	6	1	4	3		3	0.63	1	1	12	0		0	0.14	3	2	9	0		0	0.40	1	1	12	0		0	0.14
GFE 0.01	1	3	10	0	14	3	0.17	6	1	4	3	14	3	0.63	1	1	12	0	14	0	0.14	3	2	9	0	14	0	0.40	1	1	12	0	14	0	0.14
GFE 0.05	1	3	10	0	14	3	0.17	6	1	4	3	14	3	0.63	1	1	12	0	14	0	0.14	3	2	9	0	14	0	0.40	1	1	12	0	13	0	0.14
GFE 0.1	1	2	11	0	13	2	0.15	6	1	4	3	13	3	0.63	1	0	13	0	13	0	0.13	3	2	9	0	13	0	0.40	1	1	12	0	12	0	0.14
GFE 0.25	1	2	11	0	11	2	0.15	7	1	4	2	11	2	0.70	1	1	12	0	11	0	0.14	3	2	9	0	11	0	0.40	1	1	12	0	10	0	0.14
GFE 0.5	1	1	12	0	9	1	0.14	8	1	4	1	9	1	0.76	1	2	11	0	9	0	0.15	3	2	9	0	9	0	0.40	1	1	12	0	9	0	0.14
GFE 1	1	1	12	0	7	1	0.14	8	1	4	1	7	1	0.76	1	0	13	0	7	0	0.13	3	2	9	0	7	0	0.40	1	0	13	0	7	0	0.13
Cut-off: Sens + Spec																																			
Pool	1	5	8	0		0	0.20	3	2	3	6		0	0.40	0	8	5	1		0	–	0	10	1	3		0	–	0	10	3	1		0	–
ML	1	8	5	0		3	0.29	4	1	4	5		3	0.47	0	8	5	1		0	–	0	8	3	3		0	–	1	6	7	0		0	0.22
GFE 0.01	1	8	5	0	14	3	0.29	4	1	4	5	14	3	0.47	0	8	5	1	14	0	–	0	8	3	3	14	0	–	1	6	7	0	14	0	0.22
GFE 0.05	1	8	5	0	14	3	0.29	4	1	4	5	14	3	0.47	0	8	5	1	14	0	–	0	8	3	3	14	0	–	1	7	6	0	13	0	0.25
GFE 0.1	1	8	5	0	13	2	0.29	3	2	3	6	13	3	0.40	0	8	5	1	13	0	–	0	8	3	3	13	0	–	1	6	7	0	12	0	0.22
GFE 0.25	0	9	4	1	11	2	–	7	1	4	2	11	2	0.70	0	8	5	1	11	0	–	0	8	3	3	11	0	–	0	10	3	1	10	0	–
GFE 0.5	0	5	8	1	9	1	–	2	1	4	7	9	1	0.27	0	10	3	1	9	0	–	0	8	3	3	9	0	–	0	9	4	1	9	0	–
GFE 1	1	6	7	0	7	1	0.22	4	1	4	5	7	1	0.47	0	11	2	1	7	0	–	0	9	2	3	7	0	–	0	7	6	1	7	0	–
Cut-off: F1																																			
Pool	0	11	2	1		0	–	1	3	2	8		0	0.17	0	12	1	1		0	–	0	11	0	3		0	–	0	13	0	1		0	–
ML	1	8	5	0		3	0.29	1	5	0	8		3	0.20	0	11	2	1		0	–	0	10	1	3		0	–	0	13	0	1		0	–
GFE 0.01	1	8	5	0	14	3	0.29	1	5	0	8	14	3	0.20	0	11	2	1	14	0	–	0	10	1	3	14	0	–	0	13	0	1	14	0	–
GFE 0.05	1	8	5	0	14	3	0.29	1	5	0	8	14	3	0.20	0	11	2	1	14	0	–	0	10	1	3	14	0	–	0	13	0	1	13	0	–
GFE 0.1	1	8	5	0	13	2	0.29	0	5	0	9	13	3	–	0	11	2	1	13	0	–	0	10	1	3	13	0	–	0	13	0	1	12	0	–
GFE 0.25	0	13	0	1	11	2	-	2	3	2	7	11	2	0.31	0	11	2	1	11	0	–	0	10	1	3	11	0	–	0	12	1	1	10	0	–
GFE 0.5	0	13	0	1	9	1	-	0	5	0	9	9	1	–	0	12	1	1	9	0	–	0	10	1	3	9	0	–	0	13	0	1	9	0	–
GFE 1	0	13	0	1	7	1	–	0	5	0	9	7	1	–	0	12	1	1	7	0	–	0	10	1	3	7	0	–	0	13	0	1	7	0	–

"TP" true positives, "TN" true negatives, "FP" false positives, "FN" false negatives. *K* reports the number of groups for individuals obtained in the first step for GFE . "CS" is the number of countries removed due to complete separation. "F1" is the out-of-sample F1-score. Estimators: "Pooled" is the pooled model, "FE" is standard FE estimator, "GFE" is standard GFE with different values of γ_k parameter. "Cut-off" indicates the metric optimized on the training set to get the cut-off for fitted probabilities

Table 10 Banking crises forecasts: Model (1) with $x_{i,t-2}$, definition of Baron et al. (2021)

	2007							2008							2010							2011							2016						
	TP	TN	FP	FN	K	CS	F1	TP	TN	FP	FN	K	CS	F1	TP	TN	FP	FN	K	CS	F1	TP	TN	FP	FN	K	CS	F1	TP	TN	FP	FN	K	CS	F1
Cut-off: TP																																			
Pooled	1	0	13	0		0	0.13	9	1	4	0		0	0.82	0	8	5	1		0		3	1	10	0		0	0.38	1	0	13	0		0	0.13
ML	1	3	10	0		3	0.17	6	1	4	3		3	0.63	0	8	5	1		0	–	3	1	10	0		0	0.38	1	0	13	0		0	0.13
GFE 0.01	1	3	10	0	14	3	0.17	6	1	4	3	14	3	0.63	0	8	5	1	14	0	–	3	1	10	0	14	0	0.38	1	0	13	0	14	0	0.13
GFE 0.05	1	3	10	0	14	3	0.17	6	1	4	3	14	3	0.63	0	8	5	1	14	0	–	3	1	10	0	14	0	0.38	1	0	13	0	13	0	0.13
GFE 0.1	1	3	10	0	13	3	0.17	7	1	4	2	13	2	0.70	0	8	5	1	13	0	–	3	1	10	0	13	0	0.38	1	0	13	0	12	0	0.13
GFE 0.25	1	2	11	0	11	2	0.15	7	1	4	2	11	2	0.70	0	9	4	1	11	0	–	3	1	10	0	11	0	0.38	1	0	13	0	11	0	0.13
GFE 0.5	1	1	12	0	9	1	0.14	8	1	4	1	9	1	0.76	0	9	4	1	9	0	–	3	1	10	0	9	0	0.38	1	0	13	0	9	0	0.13
GFE 1	1	1	12	0	7	1	0.14	8	1	4	1	7	1	0.76	0	8	5	1	7	0	–	3	1	10	0	7	0	0.38	1	0	13	0	7	0	0.13
Cut-off: Sens + Spec																																			
Pooled	1	4	9	0		0	0.18	3	2	3	6		0	0.40	0	13	0	1		0	–	0	8	3	3		0	–	1	7	6	0		0	0.25
ML	1	3	10	0		3	0.17	5	1	4	4		3	0.56	0	13	0	1		0	–	0	9	2	3		0	–	1	9	4	0		0	0.33
GFE 0.01	1	3	10	0	14	3	0.17	5	1	4	4	14	3	0.56	0	13	0	1	14	0	–	0	9	2	3	14	0	–	1	9	4	0	14	0	0.33
GFE 0.05	1	3	10	0	14	3	0.17	5	1	4	4	14	3	0.56	0	13	0	1	14	0	–	0	9	2	3	14	0		0	10	3	1	13	0	–
GFE 0.1	1	3	10	0	13	3	0.17	7	1	4	2	13	2	0.70	0	13	0	1	13	0	–	0	9	2	3	13	0	–	0	9	4	1	12	0	–
GFE 0.25	1	2	11	0	11	2	0.15	7	1	4	2	11	2	0.70	0	13	0	1	11	0	–	0	8	3	3	11	0	–	0	9	4	1	11	0	–
GFE 0.5	1	2	11	0	9	1	0.15	7	1	4	2	9	1	0.70	0	13	0	1	9	0	–	0	10	1	3	9	0	–	0	9	4	1	9	0	–
GFE 1	1	3	10	0	7	1	0.17	3	2	3	6	7	1	0.40	0	13	0	1	7	0	–	0	10	1	3	7	0	–	0	9	4	1	7	0	–
Cut-off: F1																																			
Pooled	0	13	0	1		0	–	0	5	0	9		0	–	0	13	0	1		0		0	10	1	3		0	–	0	12	1	1		0	–
ML	0	12	1	1		3	–	1	5	0	8		3	0.20	0	13	0	1		0	–	0	10	1	3		0	–	0	12	1	1		0	–
GFE 0.01	0	12	1	1	14	3	–	1	5	0	8	14	3	0.20	0	13	0	1	14	0	–	0	10	1	3	14	0	–	0	12	1	1	14	0	–
GFE 0.05	0	12	1	1	14	3	–	1	5	0	8	14	3	0.20	0	13	0	1	14	0	–	0	10	1	3	14	0	–	0	11	2	1	13	0	–
GFE 0.1	0	12	1	1	13	3	–	2	5	0	7	13	2	0.36	0	13	0	1	13	0	–	0	10	1	3	13	0	–	0	11	2	1	12	0	–
GFE 0.25	0	13	0	1	11	2	–	2	5	0	7	11	2	0.36	0	13	0	1	11	0	–	0	10	1	3	11	0	–	0	10	3	1	11	0	–
GFE 0.5	0	13	0	1	9	1	–	1	5	0	8	9	1	0.20	0	13	0	1	9	0	–	0	10	1	3	9	0	–	0	12	1	1	9	0	–
GFE 1	0	13	0	1	7	1	–	0	5	0	9	7	1	–	0	13	0	1	7	0	–	0	10	1	3	7	0	–	0	11	2	1	7	0	–

"TP" true positives, "TN" true negatives, "FP" false positives, "FN" false negatives. K reports the number of groups for individuals obtained in the first step for GFE . "CS" is the number of countries removed due to complete separation. "F1" is the out-of-sample F1-score. Estimators: "Pooled" is the pooled model, "FE" is standard FE estimator, "GFE" is standard GFE with different values of γ_k parameter. "Cut-off" indicates the metric optimized on the training set to get the cut-off for fitted probabilities

they yield more balanced results when the cut-off is chosen as the in-sample sum of specificity and sensitivity. Optimizing the cut-off with respect to the in-sample true positive rate represents an aggressive strategy and results in a low number of true negatives; in this context, such an approach is advisable when the policy-maker holds a strong prior belief about the likelihood of critical events.

Logistic ridge offers the best predictive performance, balancing the number of correctly predicted crises with the number of true negatives: this is not surprising, since it induces a regularization which is the closest in the spirit to the one introduced by the GFE procedure. Among tree-based models, decision tree and random forest go in different directions. In decision tree, we adopt a different weighting scheme for true positives and true negatives where correctly predicted crises are weighted 100 times more than correctly predicted non-crises: the weighting solution leads to a constant performance for different cut-offs and it is suggested for a conservative priors of the policy-maker. With respect to decision tree, random forest manages to forecast a greater number of true positives, at the cost of an increased number of false alarms. Support vector machine is also a valuable choice due to its ability to offer meaningful predictions when the selected cut-off is the in-sample F1-score. Furthermore, this algorithm offers a balanced output when the cut-off is the sum of sensitivity and specificity. Finally, K-nearest neighbors seem not to be the right choice for this empirical application due to its poor ability to predict rare events for each selected cut-off (Table 11).

7 Conclusions

This chapter presents a new approach for the regularized estimation of models with rare events in the presence of severe complete separation, using the GFE method. We demonstrate its effectiveness in an application involving historical data on rare financial crises.

First, we focus on the identification of early warnings of financial turmoil and show that the GFE approach delivers reliable estimates of distress indicators. The findings are consistent with the leading literature on the determinants of financial crises (Jordà et al., 2017; Schularick & Taylor, 2012), highlighting the role of rapid credit growth to firms and individuals. Moreover, GFE relies on a more parsimonious estimation compared to the fixed effects approach, due to the reduced number of unobserved heterogeneity parameters being estimated.

Then, we apply the GFE estimator to forecasting and evaluate its performance in predicting financial crises: we show that GFE performs as well as, or better than, both FE models and MLAs. As an additional result, we show that the aforementioned advantages of the GFE approach are robust to different definitions of crisis, such as that in Baron et al. (2021). However, the GFE approach has some limitations. First, it is sensitive to the choice of the number of groups, and the parameter γ_k plays a crucial role in this decision. Additionally, the computational burden can be significant for larger datasets, due to the iterative clustering procedure required to comply with the

Table 11 Banking crises forecasts: machine learning algorithms

	2007					2008				
	TP	TN	FP	FN	F1	TP	TN	FP	FN	F1
Cut-off: TP										
Random Forest	2	0	16	0	0.2	11	0	7	0	0.759
Decision Trees	0	15	1	2	–	3	4	3	8	0.353
SVM	2	2	14	0	0.222	11	0	7	0	0.759
K-NN	0	15	1	2	–	0	5	2	11	–
RIDGE	2	4	12	0	0.25	9	1	6	2	0.692
Cut-off: Se + Sp										
Random forest	1	15	1	1	0.5	6	0	7	5	0.5
Decision trees	0	15	1	2	–	3	4	3	8	0.353
SVM	1	14	2	1	0.4	4	2	5	7	0.4
K-NN	0	15	1	2	–	0	5	2	11	–
RIDGE	1	12	4	1	0.286	5	3	4	6	0.5
Cut-off: F1-score										
Random forest	0	16	0	2	–	0	6	1	11	–
Decision trees	0	15	1	2	–	3	4	3	8	0.353
SVM	0	16	0	2	–	11	0	7	0	0.759
K-NN	0	16	0	2	–	0	6	1	11	–
RIDGE	0	16	0	2	–	0	7	0	11	–

Random Forest: 2,000 trees, minimum node size of 20 observations, and number of variables considered at each split selected via cross-validation. Decision Trees: minimum number of observations required for a split set to 200, pruning parameter equal to 0.001, maximum tree depth set to 30, and the cost of missing a crisis assigned a weight 100 times higher than that of a false alarm. SVM: radial basis function kernel, with cost and gamma parameters selected through cross-validation. K-NN: the number of neighbors (k) selected via cross-validation and set to 20. Ridge: regularization parameter (λ) selected via cross-validation

optimal group selection rule. The implementation of solutions in order to tackle these issues is left for future research.

Acknowledgement We gratefully acknowledge Gaetano Iervolino for his valuable collaboration in the development and implementation of the data analyses presented in this volume.

References

Aikman, D., Haldane, A. G., & Nelson, B. D. (2015). Curbing the credit cycle. *The Economic Journal, 125*, 1072–1109.

Albert, A., & Anderson, J. A. (1984). On the existence of maximum likelihood estimates in logistic regression models. *Biometrika, 71*, 1–10.

Alessi, L., & Detken, C. (2018). Identifying excessive credit growth and leverage. *Journal of Financial Stability, 35*, 215–225.

Aliber, R. Z., Kindleberger, C. P., & McCauley, R. N. (2023). *Manias, panics, and crashes: A history of financial crises* (8th ed.). Palgrave Macmillan.

Allaj, E., & Sanfelici, S. (2023). Early warning systems for identifying financial instability. *International Journal of Forecasting, 39*, 1777–1803.

Baron, M., Verner, E., & Xiong, W. (2021). Banking crises without panics. *The Quarterly Journal of Economics, 136*, 51–113.

Bluwstein, K., Buckmann, M., Joseph, A., Kapadia, S., & Şimşek, Ö. (2023). Credit growth, the yield curve and financial crisis prediction: Evidence from a machine learning approach. *Journal of International Economics, 145*, Article 103773.

Bonhomme, S., & Manresa, E. (2015). Grouped patterns of heterogeneity in panel data. *Econometrica,83*, 1147–1184.

Bonhomme, S., Lamadon, T., & Manresa, E. (2022). Discretizing unobserved heterogeneity. *Econometrica, 90*, 625–643.

Bordo, M., Eichengreen, B., Klingebiel, D., & Martinez-Peria, M. S. (2001). Is the crisis problem growing more severe? *Economic Policy,16*, 52–82.

Borio, C., Gambacorta, L., & Hofmann, B. (2017). The influence of monetary policy on bank profitability. *International Finance, 20*, 48–63.

Borio, C., & Lowe, P. (2002). Asset prices, financial and monetary stability: exploring the nexus. BIS Working Paper, 114, Bank for International Settlements.

Caggiano, G., Calice, P., Leonida, L., & Kapetanios, G. (2016). Comparing logit-based early warning systems: Does the duration of systemic banking crises matter? *Journal of Empirical Finance, 37*, 104–116.

Casabianca, E. J., Catalano, M., Forni, L., Giarda, E., & Passeri, S. (2022). A machine learning approach to rank the determinants of banking crises over time and across countries. *Journal of International Money and Finance, 129*, Article 102739.

Davis, E. P., & Karim, D. (2008). Comparing early warning systems for banking crises. *Journal of Financial Stability, 4*, 89–120.

Dawood, M., Horsewood, N., & Strobel, F. (2017). Predicting sovereign debt crises: An early warning system approach. *Journal of Financial Stability, 28*, 16–28.

Demirgüç-Kunt, A., & Detragiache, E. (1998). The determinants of banking crises in developing and developed countries. *IMF Staff Papers, 45*, 81–109.

Demirgüç-Kunt, A., & Detragiache, E. (2005). *Cross-country empirical studies of systemic bank distress: A survey. IMF Working Paper: Cross-Country Empirical Studies of Systemic Bank Distress-A Survey* (Vol. 5, pp. 1–33)

Dhaene, G. and K. Jochmans (2015). Split-panel jackknife estimation of fixed-effect models. *The Review of Economic Studies 82*(3 (292)), 991–1030.

Filippopoulou, C., Galariotis, E., & Spyrou, S. (2020). An early warning system for predicting systemic banking crises in the eurozone: A logit regression approach. *Journal of Economic Behavior and Organization, 172*, 344–363.

Gogas, P., Papadimitriou, T., & Agrapetidou, A. (2018). Forecasting bank failures and stress testing: A machine learning approach. *International Journal of Forecasting, 34*, 440–455.

Greenwood, R., Hanson, S. G., Shleifer, A., & Sørensen, J. A. (2022). Predictable financial crises. *The Journal of Finance, 77*, 863–921.

Hahn, J., & Kuersteiner, G. (2011). Bias reduction for dynamic nonlinear panel models with fixed effects. *Econometric Theory, 27*, 1152–1191.

Jordà, Ò., Schularick, M., & Taylor, A. M. (2017). Macrofinancial history and the new business cycle facts. *NBER Macroeconomics Annual, 31*, 213–263.

Kaminsky, G. L., & Reinhart, C. M. (1999). The twin crises: the causes of banking and balance-of-payments problems. *American Economic Review, 89*, 473–500.

Laeven, L., & Valencia, F. (2018). Systemic banking crises revisited. IMF Working Paper 2018/206, International Monetary Fund.

Lancaster, T. (2000). The incidental parameter problem since 1948. *Journal of Econometrics, 95*, 391–413.

Lang, J. H., Peltonen, T., & Sarlin, P. (2018). A framework for early-warning modeling with an application to banks Number 2182. ECB working paper.

Li, H., Lindsay, B. G., & Waterman, R. P. (2003). Efficiency of projected score methods in rectangular array asymptotics. *Journal of the Royal Statistical Society: Series B (Statistical Methodology), 65*, 191–208.

Nazareth, N., & Reddy, Y. V. R. (2023). Financial applications of machine learning: A literature review. *Expert Systems with Applications, 219*, Article 119640.

Neyman, J., & Scott, E. L. (1948). Consistent estimates based on partially consistent observations. *Econometrica,* 1–32.

Pigini, C. (2021). Penalized maximum likelihood estimation of logit-based early warning systems. *International Journal of Forecasting, 37*, 1156–1172.

Pigini, C., Pionati, A., & Valentini, F. (2025). Grouped fixed effects regularization for binary choice models. arXiv:2502.06446

R Core Team. (2025). *R: A language and environment for statistical computing*. R Foundation for Statistical Computing.

Reinhart, C. M., & Rogoff, K. S. (2009). *This time is different: Eight centuries of financial folly (Reprint paperback* (2011th ed.). Princeton University Press.

Schularick, M., & Taylor, A. M. (2012). Credit booms gone bust: monetary policy, leverage cycles, and financial crises, 1870–2008. *American Economic Review, 102*, 1029–1061.

Shen, C.-H., & Hsu, H.-H. (2022). The determinants of asian banking crises—application of the panel threshold logit model. *International Review of Finance, 22*, 248–277.

Taylor, J. B. (2009). The financial crisis and the policy responses: An empirical analysis of what went wrong. National Bureau of Economic Research. NBER Working Paper 14631.

Tölö, E. (2020). Predicting systemic financial crises with recurrent neural networks. *Journal of Financial Stability, 49*, Article 100746.

A Bayesian Student's t-Hidden Markov Model Approach for Cryptocurrencies Time Series

Romina Peruilh Bagolini, Daniele Tancini, and Silvia Pandolfi

Abstract Hidden Markov models are a widely used class of discrete latent variable models that capture temporal dependence in time series and longitudinal data. However, the commonly adopted Gaussian assumption for the conditional distribution of the responses may be unsuitable for heavy-tailed or extreme-value data. To address this limitation, we propose a multivariate Student's t-hidden Markov model within a Bayesian framework. The Bayesian setting yields predictive distributions, providing a probabilistic interpretation of latent state forecasts and a direct quantification of uncertainty. A data augmented scheme further facilitates the dynamic classification of observations into latent states. We apply the methodology to daily log-returns of Bitcoin, Ethereum, and Ripple. The model identifies latent states related to different risk levels, offering crucial insights for financial risk assessment and early warning systems. The results demonstrate the advantages of the Student's t specification in capturing complex dynamics in financial applications.

Keywords Bayesian inference · Data augmentation · Early warnings · Predictive distributions · Time series

---

Supplementary Information The online version contains supplementary material available at https://doi.org/10.1007/978-3-032-24141-2_5.

R. Peruilh Bagolini (✉) · D. Tancini · S. Pandolfi
University of Perugia, Perugia, Italy
e-mail: romina.peruilh@dottorandi.unipg.it

D. Tancini
e-mail: daniele.tancini@unipg.it

S. Pandolfi
e-mail: silvia.pandolfi@unipg.it

F. Bartolucci et al. (eds.), *Models for Longitudinal Data with Applications to Early Warning Systems*, Springer Proceedings in Mathematics & Statistics 553,
https://doi.org/10.1007/978-3-032-24141-2_5

1 Introduction

Hidden Markov models (HMMs; Bartolucci et al., 2013; Zucchini et al., 2016) are discrete latent variable models that extend finite mixture models (McLachlan & Peel, 2000) accounting for temporal dependencies in time series and longitudinal data. They have been widely applied to sequential unsupervised clustering, which aims to obtain a partition of the data without relying on reference labels. The applications of HMMs span from speech recognition (Seara Vieira et al., 2016; Young & Woodland, 1994) to gene expression dynamics (Zeng & Garcia-Frias, 2006), driving behavior (Yao et al., 2021), and finance (Langrock et al., 2012; Maruotti et al., 2019). Often, when HMMs are applied to multivariate data, the distribution of the response variables conditional on the latent process is assumed to be Gaussian. This choice reflects the simplicity and adaptability of the normal distribution in practical implementations. However, the Gaussian assumption may be inadequate for heavy-tailed data; as observed in the non-dynamic case, Gaussian mixture models tend to overestimate the number of components when approximating heavy-tailed distributions (Chatzis et al., 2008). Furthermore, atypical observations within the components of a normal mixture model can affect the estimates of the component means and covariances (Peel & McLachlan, 2000). In such cases, the inferred latent structure may fail to accurately represent the true underlying structure of the data.

An alternative to the Gaussian HMM is offered by the Student's t-hidden Markov model (SHMM; Chatzis et al., 2008; Cortese et al., 2024; Peel & McLachlan, 2000), which replaces the Gaussian distribution with a t-distribution in modeling the conditional responses. The Student's t-distribution is characterized by heavier and longer tails compared to the Gaussian distribution (Archambeau & Verleysen, 2007; Nguyen & Wu, 2013; Zhang et al., 2013). Previous studies have exploited SHMMs to improve the robustness in the presence of outliers (Chatzis & Kosmopoulos, 2011; Chatzis et al., 2008; Zhang et al., 2013; Zheng et al., 2017), showing their effectiveness as an alternative to modeling and clustering heavy-tailed data. For the use of the t-distribution in the HMM framework, see also Bernardi et al. (2017), while for alternative heavy-tailed distributions see Maruotti et al. (2019).

In this study, we propose a multivariate SHMM approach for modeling scenarios where extreme events occur more frequently than would be expected under a Gaussian distribution. Our approach is designed to handle heavy-tailed data, with a specific focus on predicting latent states in risky processes, as a strategy to generate early warning signals of the process's risk state. We formulate the SHMM within a Bayesian framework, as an alternative to the classical maximum likelihood estimation approach (Baum et al., 1970; Dempster et al., 1977). The Bayesian formulation offers significant advantages for forecasting tasks, particularly by enabling the computation of moments, such as the expected value, variance, and credible sets by considering the full predictive distribution of the data. This allows for a probabilistic interpretation of the model outputs and provides a measure of their associated uncertainty; see Frühwirth-Schnatter (2006) for an overview of Bayesian approaches for HMMs. In particular, we follow a data augmented approach in which the distribution of the model is augmented by the latent process (Tanner & Wong, 1987). This approach enables direct estimation of the latent variables, which in turn allows for a straightforward classification of the observations into latent states (or regimes).

The proposed SHMM approach is employed for temporal clustering of cryptocurrency data, aiming to identify different risk levels on the basis of the estimated hidden process underlying the log-returns of the assets. Furthermore, the Bayesian approach allows us to define the predictive distribution, which can be used to compute various risk measures, such as Value-at-Risk (Morgan, 1996) and the Expected Shortfall (Artzner et al., 1999). We focus on modeling three major capitalized cryptocurrencies: Bitcoin (BTC), Ethereum (ETH), and Ripple (XRP), using daily log-return data over a five-year period, from July 1, 2020 to July 1, 2025. Historical price data are obtained from Yahoo Finance. The code employed for the simulations and for developing the application has been implemented for the software R (R Core Team, 2025).

The remainder of this chapter is organized as follows. Section 2 presents the Bayesian general formulation for HMMs and introduces the class of SHMMs based on a multivariate t-distribution. The data augmented estimation approach and related issues such as model selection, label switching, and forecasting are discussed in Sect. 3. Section 4 presents an overview of the cryptocurrency data, along with the results of the application of the proposed model, focusing on the clustering outcomes and their implications in the context of financial risk measures. Section 5 concludes with final remarks.

2 Bayesian Hidden Markov Models

In the following section, we present the general formulation for HMMs in a Bayesian setting and we introduce the extension of the model based on the t-distribution.

2.1 *General Model Formulation*

The standard HMM formulation provides the baseline structure to build the SHMM version. The HMM is formalized by the sequence $\{(\boldsymbol{Y}_t, U_t)\}_{t=1}^{T}$, where $\boldsymbol{Y}_t \in \mathcal{Y} \subseteq \mathbb{R}^d$ is observable and $U_t \in \mathcal{U} = \{1, \ldots, k\}$ is a latent process that follows a first-order time-homogeneous Markov model. The main assumption of the model is

$$\boldsymbol{Y}_t | U_t = u \sim \mathcal{L}_u, \quad u = 1, \ldots, k, \tag{1}$$

where $\mathcal{L}_u$ is a general probability law which depends on u, with $u = 1, \ldots, k$. Let $\boldsymbol{\theta}$ be the collection of parameters based on the assumed distribution for the observable variables, and let $\boldsymbol{\pi}$ and $\boldsymbol{\Pi}$ be the vector of the initial probabilities and the transition probability matrix, respectively, which characterize the Markov process. These are defined as follows:

$$\boldsymbol{\pi} = [\pi_1, \ldots, \pi_k]',$$

where $\pi_u = p(U_1 = u)$ for $u = 1, \ldots, k$, while

$$\boldsymbol{\Pi} = \begin{bmatrix} \pi_{1,1} & \cdots & \pi_{1,k} \\ \vdots & \ddots & \vdots \\ \pi_{k,1} & \cdots & \pi_{k,k} \end{bmatrix},$$

where $\pi_{u,v} = p(U_t = v|U_{t-1} = u)$ for $u, v = 1, \ldots, k$ and $t = 2, \ldots, T$.

Assuming local independence of the observable variables given the latent variables, and independence *a priori* between the parameters, the augmented posterior distribution for the model parameters is written as

$$p(\boldsymbol{\theta}, \boldsymbol{\pi}, \boldsymbol{\Pi}, \boldsymbol{u}|\boldsymbol{y}) \propto p(\boldsymbol{y}|\boldsymbol{u}, \boldsymbol{\theta})p(\boldsymbol{u}|\boldsymbol{\pi}, \boldsymbol{\Pi})p(\boldsymbol{\theta})p(\boldsymbol{\pi})p(\boldsymbol{\Pi}), \tag{2}$$

where $\boldsymbol{u} = \{u_1, \ldots, u_T\}$ and $\boldsymbol{y} = \{\boldsymbol{y}_1, \ldots, \boldsymbol{y}_T\}$. In Eq. (2) $p(\boldsymbol{y}|\boldsymbol{u}, \boldsymbol{\theta})$ denotes the conditional distribution of the responses given the latent variables, $p(\boldsymbol{u}|\boldsymbol{\pi}, \boldsymbol{\Pi})$ is the distribution of the Markov process, while the distributions $p(\boldsymbol{\theta})$, $p(\boldsymbol{\pi})$ and $p(\boldsymbol{\Pi})$ are the prior distributions. The augmented posterior distribution follows the idea proposed in Tanner and Wong (1987), allowing to generate the latent variable U_t within the Markov chain Monte Carlo (MCMC) scheme. This approach has two main advantages: the first one is that it does not require marginalizing out the latent variables in the model, and the second is that it is possible to obtain the prediction of the unobservable states without requiring any decoding procedures. In addition, the augmented approach gives better results in terms of autocorrelation and effective sample size of the chains compared to the marginal likelihood approach, as discussed in Tancini et al. (2024).

2.2 Student's t-Hidden Markov Model

SHMMs extend HMMs by replacing the Gaussian state-dependent distributions with Student's t-distributions. In this setting, instead of directly using a multivariate t-distribution, which does not admit closed-form expressions for the full conditional distributions required for Gibbs sampling, we consider a hierarchical formulation based on a multivariate Gaussian scale-mixture model (Lin et al., 2004; Peel & McLachlan, 2000).

In detail, let $\boldsymbol{Y}_t \sim \mathcal{N}_d(\boldsymbol{\mu}, \boldsymbol{\Sigma}/\delta_t)$ and let $\delta_t \sim \mathcal{G}(\nu/2, \nu/2)$, where $\mathcal{N}_d(\cdot, \cdot)$ denotes a (d-variate) normal distribution and $\mathcal{G}(\cdot, \cdot)$ a Gamma distribution. It is known that

$$\int_{\mathbb{R}_+} \phi_d(\boldsymbol{y}_t; \boldsymbol{\mu}, \boldsymbol{\Sigma}/\delta_t)\psi(\delta_t; \nu/2, \nu/2)d\delta_t,$$

defines a multivariate t-distribution, denoted by $\mathcal{T}_d(\boldsymbol{\mu}, \boldsymbol{\Sigma}, \nu)$, with location $\boldsymbol{\mu}$, scale matrix $\boldsymbol{\Sigma}$, and degrees of freedom ν. In the above expression, $\phi_d(\cdot)$ is the probability density function of the multivariate normal distribution and $\psi(\cdot)$ is the density function of the Gamma random variable above. This assumption leads us to include an additional variable, that is δ_t, into our model, so that Eq. (1) is modified as follows:

$$\boldsymbol{Y}_t|\delta_t, U_t = u \sim \mathcal{N}_d(\boldsymbol{\mu}_u, \boldsymbol{\Sigma}_u/\delta_t), \quad t = 1, \ldots, T, \; u = 1, \ldots, k.$$

Note that this is an extension of the multivariate Student's t-mixture model described in Lin et al. (2004), where we include a constraint on the degrees of freedom of the t-distribution. Indeed, instead of considering a state-specific ν_u, $u = 1, \ldots, k$, we impose $\nu = \nu_1 = \nu_2 = \cdots = \nu_k$. This choice is motivated by the need to reduce the large number of model parameters resulting from the application and to address the potential instability in the MCMC algorithm, as well as possible problems of convergence, introduced by specific values for ν_u and δ_t. However, it is possible to relax the constraint defined for ν considering $\nu_1 \neq \nu_2 \neq \cdots \neq \nu_k$, using then Bayesian variational inference methods (Chatzis et al., 2008).

The posterior distribution defined in Eq. (2) is then written as

$$\begin{aligned} p(\boldsymbol{\mu}, \boldsymbol{\Sigma}, \boldsymbol{\pi}, \boldsymbol{\Pi}, \boldsymbol{u}, \boldsymbol{\delta}, \nu|\boldsymbol{y}) \propto\, & p(\boldsymbol{y}|\boldsymbol{u}, \boldsymbol{\mu}, \boldsymbol{\Sigma}, \boldsymbol{\delta})p(\boldsymbol{u}|\boldsymbol{\pi}, \boldsymbol{\Pi})p(\boldsymbol{\mu})p(\boldsymbol{\Sigma}) \\ & \times p(\boldsymbol{\pi})p(\boldsymbol{\Pi})p(\boldsymbol{\delta}|\nu)p(\nu), \end{aligned} \tag{3}$$

where $\boldsymbol{\mu} = \{\boldsymbol{\mu}_1, \ldots, \boldsymbol{\mu}_k\}$, $\boldsymbol{\Sigma} = \{\boldsymbol{\Sigma}_1, \ldots, \boldsymbol{\Sigma}_k\}$, $\boldsymbol{\delta} = \{\delta_1, \ldots, \delta_T\}$, whereas

$$p(\boldsymbol{y}|\boldsymbol{u}, \boldsymbol{\mu}, \boldsymbol{\Sigma}, \boldsymbol{\delta}) = \prod_{t=1}^{T} \phi_d(\boldsymbol{y}_t; \boldsymbol{\mu}_{u_t}, \boldsymbol{\Sigma}_{u_t}/\delta_t),$$

and

$$p(\boldsymbol{u}|\boldsymbol{\pi}, \boldsymbol{\Pi}) = \pi_{u_1} \prod_{t=2}^{T} \pi_{u_{t-1},u_t}.$$

Regarding the prior setting, we consider:

$$\boldsymbol{\mu}_u \sim \mathcal{N}_d(\boldsymbol{m}, \boldsymbol{V}), \quad \boldsymbol{\Sigma}_u \sim \mathcal{IW}_d(a, \boldsymbol{S}), \quad u = 1, \ldots, k,$$

where $\mathcal{IW}_d(\cdot, \cdot)$ denotes a d-dimensional Inverse-Wishart distribution with a denoting the degrees of freedom and $\boldsymbol{S}$ the scale matrix. We also assume

$$\boldsymbol{\pi} \sim \mathcal{D}(\boldsymbol{\alpha}), \quad \boldsymbol{\pi}_{u,\cdot} \sim \mathcal{D}(\boldsymbol{\alpha}), \quad u = 1, \ldots, k,$$

where $\mathcal{D}(\cdot)$ denotes a Dirichlet distribution with $\boldsymbol{\alpha} = [\alpha_1, \ldots, \alpha_k]'$ denoting a suitable vector of hyperparameters, and $\boldsymbol{\pi}_{u,\cdot}$ is the u-th row of the transition matrix $\boldsymbol{\Pi}$. Finally, we assume that

$$\delta_t|\nu \sim \mathcal{G}(\nu/2, \nu/2) \quad \text{and} \quad \nu \sim \mathcal{E}(\lambda),$$

where $\mathcal{E}(\cdot)$ denotes a modified exponential distribution with support $(2, +\infty)$ and probability density function given by

$$p(\nu|\lambda) = \lambda e^{-\lambda(\nu-2)}.$$

The modified exponential distribution with support $(2, +\infty)$ ensures a finite variance for the multivariate t-distribution. This setting of conjugate priors for $\boldsymbol{\mu}$ and $\boldsymbol{\Sigma}$ enables closed-form expressions for the full conditional distributions. Notice that it is also possible to consider a hierarchical model, following a similar approach for the Gaussian responses as described in Frühwirth-Schnatter et al. (2019). In addition, we obtain a full conditional for δ_t, for $t = 1, \ldots, T$, while for ν a Metropolis-Hastings (MH) algorithm is necessary (Hastings, 1970; Metropolis et al., 1953).

3 Bayesian Inference

In this section, we illustrate the Bayesian inferential approach based on data augmentation. Then, we discuss issues related to model selection, label switching, and forecasting procedures.

3.1 Data Augmented Approach

Bayesian estimation for SHMMs can generally be performed following different methods. In particular, two main approaches can be considered: the first is based on Bayesian non-parametric techniques (Beal et al., 2001) where an infinite number of states is employed, while the second assumes a finite number of states. In this chapter, we focus on the second approach, using an augmented method (Tanner & Wong, 1987) via Metropolis-within-Gibbs algorithm.

In detail, let $\phi_d(\boldsymbol{y}_t; \boldsymbol{\mu}_u, \boldsymbol{\Sigma}_u/\delta_t)$ be the Gaussian density evaluated in $\boldsymbol{y}_t$, with parameters $\boldsymbol{\mu}_u$ and $\boldsymbol{\Sigma}_u$. Based on Eq. (3), the full conditionals needed for the MCMC algorithm can be derived. In particular, for the latent process $\{U_t\}_{t=1}^T$ we have that

- when $t = 2, \ldots, T-1$:

$$p(U_t = u|\cdots) \propto \pi_{u_{t-1},u}\phi_d(\boldsymbol{y}_t; \boldsymbol{\mu}_u, \boldsymbol{\Sigma}_u/\delta_t)\pi_{u,u_{t+1}};$$

- when $t = 1$:

$$p(U_1 = u|\cdots) \propto \pi_u \phi_d(\boldsymbol{y}_1; \boldsymbol{\mu}_u, \boldsymbol{\Sigma}_u/\delta_1)\pi_{u,u_2};$$

- when $t = T$:

$$p(U_T = u|\cdots) \propto \pi_{u_{T-1},u}\phi_d(\boldsymbol{y}_T; \boldsymbol{\mu}_u, \boldsymbol{\Sigma}_u/\delta_T).$$

The notation "$\cdots$" means "given all the other variables and parameters". Regarding the full conditional distribution for $\boldsymbol{\pi}$ we obtain that

$$\boldsymbol{\pi}|\cdots \sim \mathcal{D}(\boldsymbol{\alpha} + \boldsymbol{n}_{u_1}),$$

where

$$\boldsymbol{n}_{u_1} = \left[\mathbb{I}_{u_1}(1), \ldots, \mathbb{I}_{u_1}(k)\right]',$$

and $\mathbb{I}_{u_1}(u)$ is an indicator function equal to 1 if $u_1 = u$ and 0 otherwise, while for each row of $\boldsymbol{\Pi}$ we have

$$\boldsymbol{\pi}_{u,\cdot}|\cdots \sim \mathcal{D}(\boldsymbol{\alpha} + \boldsymbol{n}_{u,\cdot}),$$

with

$$\boldsymbol{n}_{u,\cdot} = \left[\sum_{t=2}^{T} \mathbb{I}_{u_{t-1},u_t}(u, 1), \ldots, \sum_{t=2}^{T} \mathbb{I}_{u_{t-1},u_t}(u, k)\right]',$$

where $\mathbb{I}_{u_{t-1},u_t}(u, v)$ is an indicator function equal to 1 if $u_{t-1} = u$ and $u_t = v$, and to 0 otherwise; see Tancini et al. (2024). For the mean vectors we have that

$$\boldsymbol{\mu}_u|\cdots \sim \mathcal{N}_d\left(\tilde{\boldsymbol{V}}_u \tilde{\boldsymbol{m}}_u, \tilde{\boldsymbol{V}}_u\right),$$

where

$$\tilde{\boldsymbol{V}}_u^{-1} = \sum_{t=1}^{T} \mathbb{I}_{u_t}(u)\delta_t \boldsymbol{\Sigma}_u^{-1} + \boldsymbol{V}^{-1},$$

and

$$\tilde{\boldsymbol{m}}_u = \boldsymbol{\Sigma}_u^{-1} \sum_{t=1}^{T} \boldsymbol{y}_t \mathbb{I}_{u_t}(u)\delta_t + \boldsymbol{V}^{-1}\boldsymbol{m}.$$

For the covariance matrices we have

$$\boldsymbol{\Sigma}_u|\cdots \sim \mathcal{IW}_d(a + n_u, \boldsymbol{S} + \tilde{\boldsymbol{S}}_u),$$

where $n_u = \sum_{t=1}^{T} \mathbb{I}_{u_t}(u)$ and

$$\tilde{\boldsymbol{S}}_u = \sum_{t=1}^{T} (\boldsymbol{y}_t - \boldsymbol{\mu}_u)(\boldsymbol{y}_t - \boldsymbol{\mu}_u)' \mathbb{I}_{u_t}(u)\delta_t.$$

Finally, for δ_t we have

$$\delta_t|\cdots \sim \mathcal{G}\left(\frac{d+\nu}{2}, \frac{\xi_t+\nu}{2}\right),$$

where $\xi_t = \sum_{u=1}^{k} \mathbb{I}_{u_t}(u)(\boldsymbol{y}_t - \boldsymbol{\mu}_u)'\boldsymbol{\Sigma}_u^{-1}(\boldsymbol{y}_t - \boldsymbol{\mu}_u)$. All the full conditional derivations are reported in Appendix A.

For ν we consider a random walk MH, which requires mapping ν from $(2, +\infty)$ to $\mathbb{R}$. To achieve this, we apply the logarithmic transformation

$$\eta = f(\nu) = \log(\nu - 2),$$

and then use as proposal

$$\tilde{\eta} = \eta + \epsilon, \quad \epsilon \sim \mathcal{N}(0, \sigma_\eta^2).$$

Notice that $\nu = f^{-1}(\eta) = e^\eta + 2$ and the acceptance probability of the MH includes also the ratio $\left|\frac{df^{-1}(\tilde{\eta})}{d\tilde{\eta}}\right| / \left|\frac{df^{-1}(\eta)}{d\eta}\right|$, due to the transformation used.

3.2 *Model Selection and Label Switching*

Selecting the number of components can be interpreted as a model selection problem. Let $\{\mathcal{M}_k : k = 1, \ldots, k_{\max}\}$ be a set of models with a different number of states, from 1 to $k_{\max}$, where $k_{\max}$ denotes the maximum number of states. In the Bayesian framework, the most common approach is based on the marginal likelihood (or evidence), which is used for the evaluation of the Bayes factor (Kass & Raftery, 1995). Unfortunately, the evidence is difficult to evaluate and Monte Carlo methods are usually required; see Llorente et al. (2023) for a review.

Alternatively, when the number of states is not fixed *a priori*, a typical approach is based on the Reversible jump MCMC (RJMCMC; Green, 1995). This method is based on creating a Markov chain over a space of variable dimensions, allowing transitions between models with different number of states, through moves designed to ensure reversibility and to preserve the correct stationary distribution. Different approaches may be also considered in the non-parametric Bayesian literature, where an infinite number of states might be included; in this setting a well-known work is Beal et al. (2001).

An alternative approach to the RJMCMC has been proposed in Stephens (2000a), where an MCMC algorithm that views the parameters of the model as a point process is used. An extension of this approach used in the context of HMMs can be found in Cappé et al. (2003).

When the number of states is fixed *a priori*, the model can be estimated for a chosen value of k, and standard information criteria can be used for model comparison. Some commonly used Bayesian model selection criteria include the Deviance Information

Criterion (Spiegelhalter et al., 2002), the Widely Applicable Information Criterion (Watanabe & Opper, 2010), and the Widely Applicable Bayesian Information Criterion (Watanabe, 2013). Recently, Vehtari et al. (2017) proposed a leave-one-out expected log pointwise predictive density (elpd_{loo}) approach, which can be evaluated using the R package **loo** (Vehtari et al., 2015), obtaining the leave-one-out cross-validation information criterion. However, this criterion may be unsuitable in our context, due to temporal dependence, requiring a modified approach. Finally, an alternative approach is the parallel sampling (Congdon, 2006), which provides an approximation of the marginal posterior probability of a model given the observed data. These probabilities can then be used to identify the best model by selecting the one with the highest value. We use this last approach to estimate the number of hidden states in the application to the cryptocurrency dataset.

As for the standard HMMs, estimation of SHMMs via MCMC methods leads to an identifiability issue known as label switching, which depends on the fact that the likelihood and the prior distributions are invariant under permutations of the state labels. This is a common problem also in mixture models, which has been addressed in several papers (Celeux et al., 2000; Stephens, 2000b). Different solutions have been proposed, such as the pivotal reordering (Marin et al., 2005), the probabilistic relabeling algorithm (Sperrin et al., 2010), and the data-based algorithm (Rodríguez & Walker, 2014). In this work we consider an identifiability constraint on the variance referred to the third variable in the covariance matrix, that is, the variance associated with the XRP crypto, assuming an ascending order among the XRP variances across the different latent states.

3.3 Bayesian Forecasting

The Bayesian statistical paradigm uses probability rules to quantify uncertainty about unknown aspects of the data-generating process. This characteristic is particularly suitable for forecasting, as uncertainty about the unknown values of future observations is naturally expressed in terms of a probability distribution; see Martin et al. (2024) for a review.

Formally, define $\boldsymbol{Y}_{T+1}$ as the one-step-ahead future random variable, and based on the assumptions of the model defined in Eq. (3), standard probability theory leads to the following expression for the predictive (or forecasting) distribution:

$$p(\boldsymbol{y}_{T+1}|\boldsymbol{y}) = \int \dots \int \sum_{\boldsymbol{u},u_{T+1}} p(\boldsymbol{y}_{T+1}|u_{T+1}, \boldsymbol{u}, \boldsymbol{\mu}, \boldsymbol{\Sigma}, \delta_{T+1}, \boldsymbol{\delta}, \boldsymbol{y}) p(u_{T+1}|\boldsymbol{u}, \boldsymbol{\pi}, \boldsymbol{\Pi}) \\ p(\delta_{T+1}|\nu) p(\boldsymbol{\mu}, \boldsymbol{\Sigma}, \boldsymbol{\pi}, \boldsymbol{\Pi}, \boldsymbol{u}, \boldsymbol{\delta}, \nu|\boldsymbol{y}) d\boldsymbol{\mu} \dots d\nu. \tag{4}$$

Based on the previous formulation we can define $p(\boldsymbol{y}_{T+h}|\boldsymbol{y})$, for an h-step-ahead future random variable. The integral in Eq. (4) is mathematically intractable, but it can be approximated using the samples of MCMC algorithm described in Sect. 3.1.

The previous predictive distribution can be used in different ways. It is possible to obtain a point prediction based on the conditional expectation, that is,

$$\mathbb{E}[\boldsymbol{Y}_{T+1}|\boldsymbol{y}] = \int \boldsymbol{y}_{T+1} p(\boldsymbol{y}_{T+1}|\boldsymbol{y}) d\boldsymbol{y}_{T+1},$$

or we can obtain the variance $\mathbb{V}[\boldsymbol{Y}_{T+1}|\boldsymbol{y}]$. Also in this case, the expectation and the variance can be approximated via Monte Carlo methods. Finally, it is possible to obtain the highest predictive region (HPR) at level $c \in [0, 1]$.

4 Application: Cryptocurrency Data

In the following section, we present the results of the proposed approach applied to the cryptocurrency dataset, with a focus on predicting the risk level of future log-returns, while accounting for the heavy-tailed nature of the distribution.

In detail, Sect. 4.1 presents the cryptocurrency dataset, analyzing some descriptive characteristics of the time series. Section 4.2 shows the main results related to model selection and MCMC diagnostics. Section 4.3 illustrates the estimation results and provides an interpretation of the risk levels associated with the latent variables. Finally, Sect. 4.4 discusses the forecasting procedure and possible applications in terms of financial risk measures.

4.1 Data on Cryptocurrency Log-Returns

Nowadays, BTC stands out as the most influential cryptocurrency. Its market capitalization has reached 2.32 trillion USD (https://coinmarketcap.com, accessed on 8 August 2025). Other important cryptocurrencies such as ETH and XRP are no exception and their prices have increased, reaching a market capitalization of 470.06 and 197.19 billion, respectively, retrieved from an open-access source (https://coinmarketcap.com, accessed on 8 August 2025).

In this application, we analyze the daily log-returns of BTC, ETH, and XRP for a period of 5 years. The log-returns are calculated using closing prices retrieved from the Yahoo Finance website from July 1, 2020 to July 1, 2025. Table 1 summarizes the main statistics for each series.

BTC presents a slightly positive average daily log-return of approximately 0, as expected, and a standard deviation of 3.1%, indicating moderate volatility in comparison to the other cryptocurrencies. The distribution of BTC returns is almost symmetric (skewness = −0.12) and highly leptokurtic (kurtosis = 6.561), highlighting that extreme events occur more frequently than under the Gaussian distribution. ETH also shows a daily mean near 0 but with higher volatility (Std. Dev. = 0.042) than BTC. Moreover, the ETH distribution is characterized by a slightly more negative

Table 1 Descriptive statistics of BTC, ETH, and XRP series observed over the period from July 1, 2020 to July 1, 2025

Crypto	Mean	Std. Dev.	Skewness	Kurtosis	Min.	Median	Max.
BTC	0.001	0.031	−0.122	6.561	−0.174	0.001	0.172
ETH	0.001	0.042	−0.246	8.058	−0.317	0.001	0.231
XRP	0.001	0.055	0.745	22.299	−0.055	0.001	0.549

asymmetry (skewness = −0.25) and higher kurtosis (8.06). XRP exhibits higher volatility (Std. Dev. = 0.055), and, unlike the other two series, is positively skewed (skewness = 0.75), suggesting that extreme positive returns were more frequent than extreme losses during the five-year period. Nevertheless, the XRP return series presents the highest kurtosis (22.30), indicating a greater probability of extreme events and thus higher risk.

Figure 1 plots the daily log-returns trend of BTC, ETH, and XRP between 2020 and 2025.

The visual representation of the series shows fluctuations around 0 and episodes of extreme volatility. The BTC series appears relatively stable, with occasional episodes of high volatility, consistently with its benchmark nature. ETH exhibits a pattern similar to BTC but with more pronounced volatility. XRP, on the other hand, is characterized by more intense gains, more frequent fluctuations, and highly concentrated volatility, particularly during 2021–2022.

4.2 Model Estimation

We estimate the proposed SHMM introduced in Sect. 2.2 on the cryptocurrency log-returns, for a varying number of regimes, using the data augmented approach described in Sect. 3.1. We use the log-returns expressed in percentage terms, that is,

$$y_t = \log\left(\frac{p_t}{p_{t-1}}\right) \times 100,$$

where p_t and p_{t-1} denote the prices at time t and $t-1$, respectively. Expressing log-returns in percentages simplifies the selection of the hyperparameters of the prior for the covariance matrix, helping to avoid overly concentrated distributions around small values. We collect the prices of the three cryptocurrencies at time t in the vector $\boldsymbol{y}_t$.

The models are estimated by fixing the number of iterations for the MCMC chains to 20,000, discarding the first 10% of the draws as burn-in. No optimal thinning is used. We fix the initial state probabilities as $\boldsymbol{\pi} = [1/k, \ldots, 1/k]'$, while for the transition matrix $\boldsymbol{\Pi}$ and the covariance matrices $\boldsymbol{\Sigma}_u$ we define the corresponding priors as described in Tancini et al. (2024). For the means, we assume $\boldsymbol{\mu}_u \sim \mathcal{N}_d(\boldsymbol{0}, 100\boldsymbol{I})$, for $u = 1, \ldots, k$, where $\boldsymbol{I}$ denotes an identity matrix.

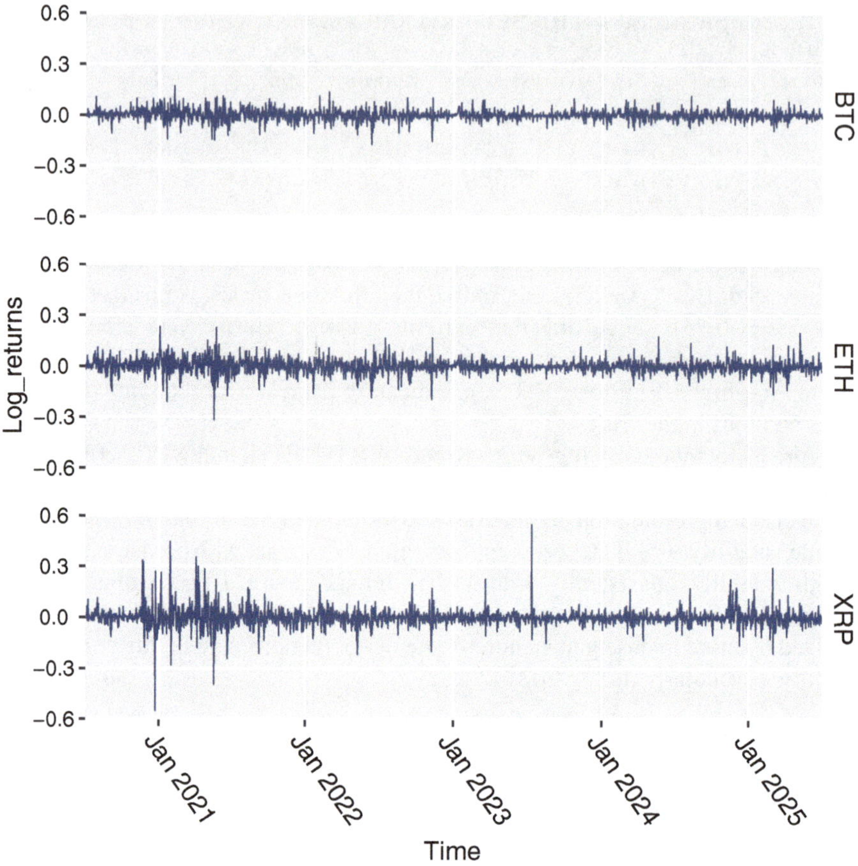

Fig. 1 Log-returns trend of BTC, ETH, and XRP series over the period from July 1, 2020 to July 1, 2025

The analysis focuses on models characterized by $k = 3, 4, 5$ states, as this range is typically considered in literature for similar applications, also for forecasting purposes; see, in particular, Giudici and Abu Hashish (2020) and Koki et al. (2022). Once the SHMMs with $k = 3, 4, 5$ are estimated, the parallel sampling approach proposed in Congdon (2006) is implemented. The corresponding posterior probabilities of the three models are: 0.429 for $k = 3$, 0.525 for $k = 4$, and 0.046 for $k = 5$. These results indicate that the most likely models are those with $k = 3$ and $k = 4$, while the model with $k = 5$ has a very low probability. Although the distinction between $k = 3$ and $k = 4$ is not well defined, the difference of 0.1 in their posterior probabilities slightly favors selecting an SHMM with $k = 4$ states.

For the convergence diagnostics of the chains, we consider the multivariate version of the effective sample size, denoted as multiESS, and the stopping rule proposed by Vats et al. (2019). The stopping rule works on the basis of a lower bound which allows us to define *a priori* the minimum multiESS required to achieve the desired tolerance and confidence levels. The use of the multiESS is related to the large number of parameters in the proposed model, for which it is difficult to analyze individual ESS values or tests. In particular, for the multiESS we consider the entire set of parameters $\boldsymbol{\mu}$, $\boldsymbol{\Sigma}$, $\boldsymbol{\Pi}$, ν, and a subset of $\boldsymbol{\delta}$, selecting δ_t at every 100 time points. This choice is driven by the computational intractability of the matrices required for the multiESS when all δ_t for $t = 1, \ldots, T$, are included. The estimated multiESS value is

$$\widehat{\text{multiESS}} = 8360.563.$$

The lower bound of the stopping rule proposed in Vats et al. (2019) suggests a minimum number of multiESS = 8221, considering the number of parameters involved, and the default values of the confidence and tolerance levels, both equal to 0.05. These values are obtained using the R package **mcmcse**.

4.3 Model Results and Interpretation

The estimated model parameters, obtained as conditional expectations with respect to the posterior distribution, are reported below

- Estimated means:

$$\hat{\boldsymbol{\mu}}_1 = [0.029, -0.057, -0.233]', \quad \hat{\boldsymbol{\mu}}_2 = [1.385, 1.872, 0.402]',$$

$$\hat{\boldsymbol{\mu}}_3 = [-0.770, -1.255, -0.289]', \quad \hat{\boldsymbol{\mu}}_4 = [0.064, 0.369, 1.461]';$$

- Estimated covariance matrices:

$$\hat{\boldsymbol{\Sigma}}_1 = \begin{bmatrix} 1.278 & 1.389 & 1.157 \\ 1.389 & 2.302 & 1.651 \\ 1.157 & 1.651 & 2.246 \end{bmatrix}, \quad \hat{\boldsymbol{\Sigma}}_2 = \begin{bmatrix} 10.157 & 8.22 & 6.986 \\ 8.222 & 14.462 & 8.675 \\ 6.986 & 8.675 & 7.438 \end{bmatrix},$$

$$\hat{\boldsymbol{\Sigma}}_3 = \begin{bmatrix} 7.101 & 7.926 & 9.032 \\ 7.926 & 11.810 & 11.910 \\ 9.032 & 11.910 & 22.802 \end{bmatrix}, \quad \hat{\boldsymbol{\Sigma}}_4 = \begin{bmatrix} 39.505 & 44.806 & 45.277 \\ 44.806 & 77.334 & 67.080 \\ 45.277 & 67.080 & 197.887 \end{bmatrix};$$

- Estimated transition probability matrix:

$$\hat{\boldsymbol{\Pi}} = \begin{bmatrix} 0.570 & 0.202 & 0.218 & 0.010 \\ 0.260 & 0.419 & 0.246 & 0.075 \\ 0.241 & 0.216 & 0.427 & 0.116 \\ 0.022 & 0.157 & 0.238 & 0.583 \end{bmatrix};$$

- Estimated degrees of freedom: $\hat{\nu} = 7.576$.

The hidden states are sorted in ascending order according to their total within-state variability, measured by the determinant of the corresponding covariance matrix. The first state is associated with the lowest overall within-state variability, while the fourth state is associated with the highest one. Consequently, these states can be interpreted in terms of volatility levels: low, medium-low, medium-high, and high volatility.

In more detail, we observe that the means of the percentage log-returns are close to 0 for state 1 (low volatility level), while for state 2 (medium-low) and state 4 (high volatility), these values are all positive. For state 3 (medium-high volatility), a negative expected conditional log-return is estimated. We also observe a consistently high level of correlation among the different cryptocurrencies across all covariance matrices, conditional on the latent states. This suggests that, regardless of the regime or market condition represented by each state, the assets tend to move together, highlighting strong co-movement dynamics within the crypto market. It is worth noting that, while in this illustrative analysis we allow for state-specific covariance matrices, assuming a constant covariance across states would enhance model parsimony, but at the expense of flexibility in capturing volatility heterogeneity.

The estimated transition probability matrix shows a moderate tendency of the process to remain in the same state over time. This persistence is more pronounced in the extreme states (states 1 and 4), whose self-transition probabilities (0.57 and 0.58, respectively) are higher than those in the intermediate states. This is an expected result, as it is related to the volatility clustering phenomenon. Similar results have been observed in the context of HMMs; see in particular Pennoni et al. (2022).

Finally, using the maximum a posteriori (MAP) rule it is possible to directly allocate units into the latent states; a graphical representation of the series according to these predicted states is reported in Fig. 2. It is worth noting that the typical clustering patterns associated with log-returns are evident in our case; in particular, the four distinct states are clearly identified and visible in the plot.

4.4 Model Forecasting

A sample from the predictive distribution $p(\boldsymbol{y}_{T+1}|\boldsymbol{y})$ can be drawn using the augmented MCMC algorithm, as outlined in Sect. 3.3. A graphical representation of the marginal predictive distribution, associated with the BTC log-returns expressed in percentage terms at $T+1$ is reported in Fig. 3.

The estimated distribution is characterized by a skewness equal to 0.318, and by a kurtosis equal to 10.436. Unlike Gaussian responses, the Student's t-distribution captures fat tails in the predictive distribution, which is particularly useful in the context of cryptocurrency data. A similar approach can be used to sample from $p(\boldsymbol{y}_{T+h}|\boldsymbol{y})$, where h defines an h-step-ahead forecast horizon. We report in Table 2 the expectation, the variance and the HPR for the predictive distribution of the BTC log-returns evaluated in percentage terms, considering 10 days ahead.

In addition, financial risk indicators can be obtained from the predictive distribution previously defined. Some of these indicators are the Value-at-Risk and

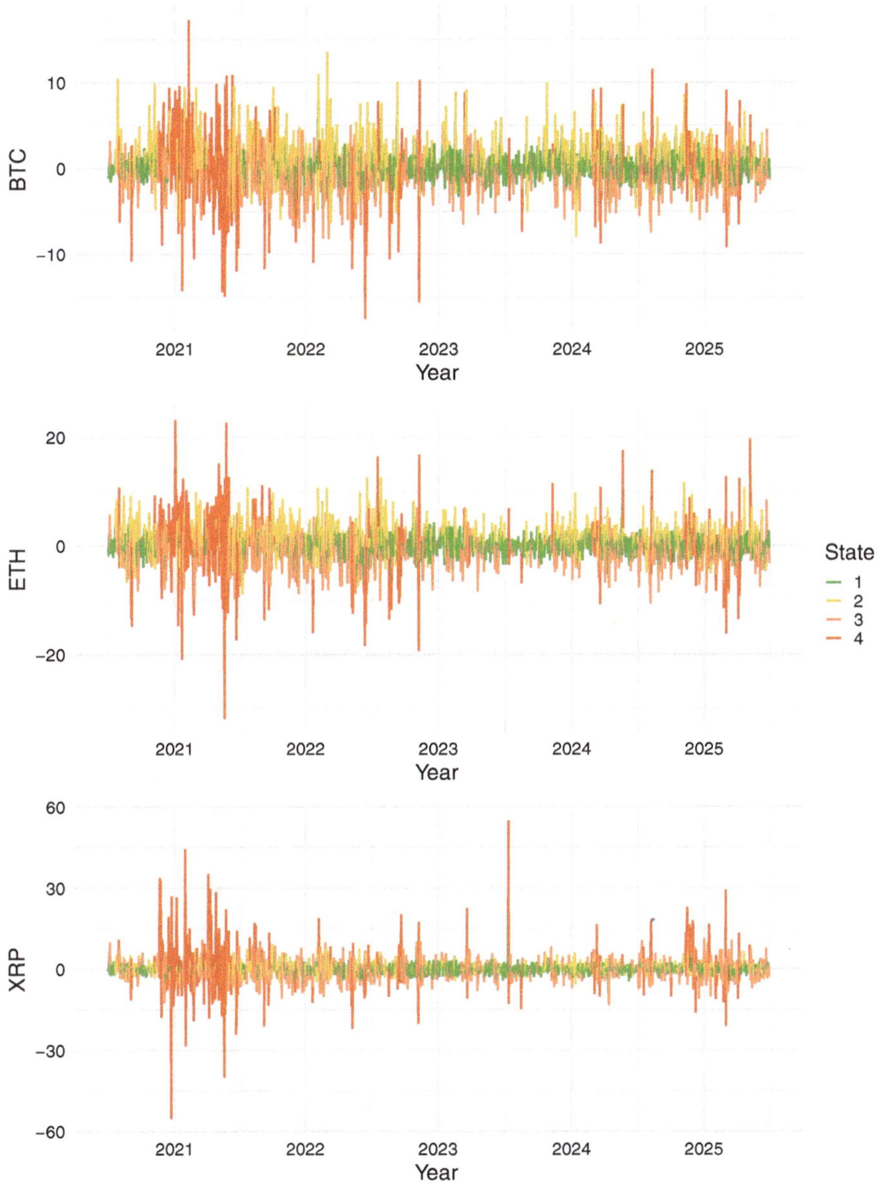

Fig. 2 Percentage log-return series for BTC, ETH, and XRP colored according to the predicted latent state over the period from July 1, 2020 to July 1, 2025

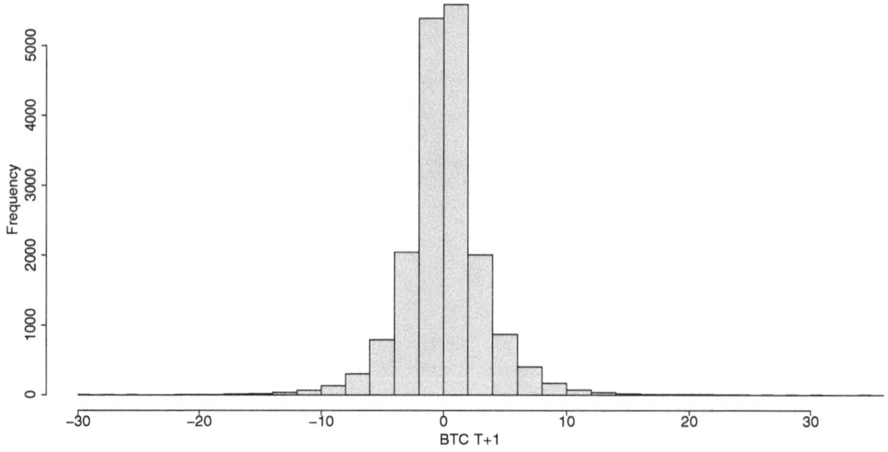

Fig. 3 Predictive distribution for the log-returns of BTC expressed in percentage terms at time $T + 1$

Table 2 Marginal expectation, variance, and HPR at level 0.95 for the predictive distribution of the BTC's log-returns evaluated in percentage terms. The notation Y_{1T+1} means the first dimension, that is BTC position, of $\boldsymbol{Y}_{T+1}$

Crypto	Time	$\mathbb{E}[Y_{1T+1}\|\boldsymbol{y}]$	$\mathbb{V}[Y_{1T+1}\|\boldsymbol{y}]$	$\text{HPR}_{0.95}$
BTC	T+1	0.098	11.335	−6.641 7.176
	T+2	0.088	13.131	−7.279 7.651
	T+3	0.110	13.760	−7.345 8.199
	T+4	0.122	14.843	−7.757 8.152
	T+5	0.090	13.945	−7.372 8.306
	T+6	0.128	14.230	−7.538 8.496
	T+7	0.181	15.014	−8.013 7.845
	T+8	0.114	14.882	−8.272 7.937
	T+9	0.074	14.606	−7.977 7.991
	T+10	0.111	15.227	−7.756 8.026

the Expected Shortfall. Moreover, from the estimated predictive distribution of the latent variables we can obtain the predicted latent states for $T + 1, \ldots, T + h$. Subsequently, these predictions can be used to construct an indicator to predict the volatility levels conditioned on the observed previous state assignment.

5 Conclusions

In this study, we consider a multivariate Student's t-hidden Markov model (SHMM) within a Bayesian framework to address the limitations of standard Gaussian hidden Markov models in modeling heavy-tailed financial time series. By replacing the Gaussian assumption with the more flexible t-distribution, our approach is effective in capturing extreme events and volatility clusters.

Our Bayesian formulation entails several advantages, including the ability to quantify uncertainty, compute the predictive distributions, and directly estimate the latent states through data augmentation. The Gibbs-within-Metropolis algorithm allows us to obtain samples from the posterior distributions, which are then used to approximate the predictive distribution.

Applying the model to daily log-returns of Bitcoin, Ethereum, and Ripple over a five-year period, we identify distinct latent regimes corresponding to varying levels of financial risk. In particular, we detect four risk levels: low, medium-low, medium-high, and high volatility. Furthermore, this approach allows us to predict the corresponding latent states for future observations. The chart could then be used to raise timely warning signals of risk in financial markets.

Overall, this work contributes to the growing literature on robust probabilistic modeling of time-series data, highlighting the potential of SHMMs for practical applications in financial risk management, particularly in high-volatility environments like the cryptocurrency market.

The proposed method can be combined with financial risk measures, such as Value-at-Risk and Expected Shortfall, allowing for a more comprehensive risk assessment framework.

Appendix A: Derivation of the Full Conditional Distributions

In this appendix we derive the full conditional distributions obtained for the model proposed. Assuming k hidden states, the full conditional for $\boldsymbol{\mu}_u$, which we define $p(\boldsymbol{\mu}_u|\cdots)$ for simplicity, is

$$
\begin{aligned}
p(\boldsymbol{\mu}_u|\cdots) &\propto \prod_{t=1}^{T} e^{-\frac{1}{2}(\boldsymbol{y}_t-\boldsymbol{\mu}_u)'\delta_t \boldsymbol{\Sigma}_u^{-1}(\boldsymbol{y}_t-\boldsymbol{\mu}_u)\mathbb{I}_{u_t}(u)} e^{-\frac{1}{2}(\boldsymbol{\mu}_u-\boldsymbol{m})'\boldsymbol{V}^{-1}(\boldsymbol{\mu}_u-\boldsymbol{m})} \qquad (5)\\
&= e^{-\frac{1}{2}\sum_t(\boldsymbol{y}_t-\boldsymbol{\mu}_u)'\delta_t \boldsymbol{\Sigma}_u^{-1}(\boldsymbol{y}_t-\boldsymbol{\mu}_u)\mathbb{I}_{u_t}(u)} e^{-\frac{1}{2}(\boldsymbol{\mu}_u-\boldsymbol{m})'\boldsymbol{V}^{-1}(\boldsymbol{\mu}_u-\boldsymbol{m})}\\
&= e^{-\frac{1}{2}\left[\sum_t \boldsymbol{y}_t'\delta_t \boldsymbol{\Sigma}_u^{-1}\boldsymbol{y}_t\mathbb{I}_{u_t}(u)-2\sum_t \boldsymbol{\mu}_u'\delta_t \boldsymbol{\Sigma}_u^{-1}\boldsymbol{y}_t\mathbb{I}_{u_t}(u)+\sum_t \boldsymbol{\mu}_u'\delta_t \boldsymbol{\Sigma}_u^{-1}\boldsymbol{\mu}_u\mathbb{I}_{u_t}(u)\right]}\\
&\quad \times e^{-\frac{1}{2}\left(\boldsymbol{\mu}_u'\boldsymbol{V}^{-1}\boldsymbol{\mu}_u-2\boldsymbol{\mu}_u'\boldsymbol{V}^{-1}\boldsymbol{m}+\boldsymbol{m}'\boldsymbol{V}^{-1}\boldsymbol{m}\right)}\\
&\propto e^{-\frac{1}{2}\left\{\boldsymbol{\mu}_u'(\sum_{t=1}^{T}\mathbb{I}_{u_t}(u)\delta_t \boldsymbol{\Sigma}_u^{-1}+\boldsymbol{V}^{-1})\boldsymbol{\mu}_u-2\boldsymbol{\mu}_u'\left[\boldsymbol{\Sigma}_u^{-1}\sum_{t=1}^{T}\boldsymbol{y}_t\mathbb{I}_{u_t}(u)\delta_t+\boldsymbol{V}^{-1}\boldsymbol{m}\right]\right\}},
\end{aligned}
$$

and from Eq. (5) it is possible to identify the Gaussian kernel.

Given k hidden states, $p(\mathbf{\Sigma}_u|\cdots)$ may be evaluated as

$$\begin{aligned} p(\mathbf{\Sigma}_u|\cdots) &\propto |\mathbf{\Sigma}_u|^{-\frac{n_u}{2}} e^{-\frac{1}{2}\sum_t (y_t-\mu_u)'\delta_t \mathbf{\Sigma}_u^{-1}(y_t-\mu_u)\mathbb{I}_{u_t}(u)} |\mathbf{\Sigma}_u|^{-\frac{a+d+1}{2}} e^{-\frac{1}{2}tr(S\mathbf{\Sigma}_u^{-1})} \\ &= |\mathbf{\Sigma}_u|^{-\frac{a+n_u+d+1}{2}} e^{-\frac{1}{2}tr\{[S+\sum_t (y_t-\mu_u)(y_t-\mu_u)'\delta_t \mathbb{I}_{u_t}(u)]\mathbf{\Sigma}_u^{-1}\}}, \end{aligned} \tag{6}$$

and from Eq. (6) it is possible to recognize the Inverse-Wishart kernel.

Given a time $t \in \{1, \ldots, T\}$, the full conditional distribution $p(\delta_t|\cdots)$ is

$$\begin{aligned} p(\delta_t|\cdots) &\propto \prod_{u=1}^{k} \left|\frac{\mathbf{\Sigma}_u}{\delta_t}\right|^{-\frac{1}{2}\mathbb{I}_{u_t}(u)} e^{-\frac{1}{2}(y_t-\mu_u)'\delta_t \mathbf{\Sigma}_u^{-1}(y_t-\mu_u)\mathbb{I}_{u_t}(u)} \\ &\quad \times \delta_t^{\frac{\nu}{2}-1} e^{-\frac{\nu}{2}\delta_t} \\ &\propto \delta_t^{\frac{d+\nu}{2}-1} e^{-\frac{\nu+\sum_{u=1}^{k}(y_t-\mu_u)'\mathbf{\Sigma}_u^{-1}(y_t-\mu_u)\mathbb{I}_{u_t}(u)}{2}\delta_t}, \end{aligned} \tag{7}$$

and from Eq. (7) it is possible to recognize the Gamma kernel.

References

Archambeau, C., & Verleysen, M. (2007). Robust Bayesian clustering. *Neural Networks, 20*, 129–138.

Artzner, P., Delbaen, F., Eber, J.-M., & Heath, D. (1999). Coherent measures of risk. *Mathematical Finance, 9*, 203–228.

Bartolucci, F., Farcomeni, A., & Pennoni, F. (2013). *Latent Markov models for longitudinal data*. Chapman and Hall/CRC.

Baum, L. E., Petrie, T., Soules, G., & Weiss, N. (1970). A maximization technique occurring in the statistical analysis of probabilistic functions of Markov chains. *The Annals of Mathematical Statistics, 41*, 164–171.

Beal, M., Ghahramani, Z., & Rasmussen, C. (2001). The infinite hidden Markov model. In *Advances in neural information processing systems* (Vol. 14).

Bernardi, M., Maruotti, A., & Petrella, L. (2017). Multiple risk measures for multivariate dynamic heavy–tailed models. *Journal of Empirical Finance, 43*, 1–32.

Cappé, O., Robert, C. P., & Rydén, T. (2003). Reversible jump, birth-and-death and more general continuous time Markov chain Monte Carlo samplers. *Journal of the Royal Statistical Society, Series B, 65*, 679–700.

Celeux, G., Hurn, M., & Robert, C. P. (2000). Computational and inferential difficulties with mixture posterior distributions. *Journal of the American Statistical Association, 95*, 957–970.

Chatzis, S. P., & Kosmopoulos, D. I. (2011). A variational Bayesian methodology for hidden Markov models utilizing Student's-t mixtures. *Pattern Recognition, 44*, 295–306.

Chatzis, S. P., Kosmopoulos, D. I., & Varvarigou, T. A. (2008). Robust sequential data modeling using an outlier tolerant hidden Markov model. *IEEE Transactions on Pattern Analysis and Machine Intelligence, 31*, 1657–1669.

Congdon, P. (2006). Bayesian model choice based on Monte Carlo estimates of posterior model probabilities. *Computational Statistics and Data Analysis, 50*, 346–357.

Cortese, F., Pennoni, F., & Bartolucci, F. (2024). Maximum likelihood estimation of multivariate regime switching Student-t copula models. *International Statistical Review, 92*, 327–354.

Dempster, A. P., Laird, N. M., & Rubin, D. B. (1977). Maximum likelihood from incomplete data via the EM algorithm. *Journal of the Royal Statistical Society, Series B, 39*, 1–38.

Frühwirth-Schnatter, S. (2006). *Finite mixture and Markov switching models*. Springer.

Frühwirth-Schnatter, S., Celeux, G., & Robert, C. P. (2019). *Handbook of mixture analysis*. CRC Press.

Giudici, P., & Abu Hashish, I. (2020). A hidden Markov model to detect regime changes in cryptoasset markets. *Quality and Reliability Engineering International, 36*, 2057–2065.

Green, P. J. (1995). Reversible jump Markov chain Monte Carlo computation and Bayesian model determination. *Biometrika, 82*, 711–732.

Hastings, W. K. (1970). Monte Carlo sampling methods using Markov chains and their applications. *Biometrika, 57*, 97–109.

Kass, R. E., & Raftery, A. E. (1995). Bayes factors. *Journal of the American Statistical Association, 90*, 773–795.

Koki, C., Leonardos, S., & Piliouras, G. (2022). Exploring the predictability of cryptocurrencies via Bayesian hidden Markov models. *Research in International Business and Finance, 59*, Article 101554.

Langrock, R., MacDonald, I. L., & Zucchini, W. (2012). Some nonstandard stochastic volatility models and their estimation using structured hidden Markov models. *Journal of Empirical Finance, 19*, 147–161.

Lin, T. I., Lee, J. C., & Ni, H. F. (2004). Bayesian analysis of mixture modelling using the multivariate t distribution. *Statistics and Computing, 14*, 119–130.

Llorente, F., Martino, L., Delgado, D., & Lopez-Santiago, J. (2023). Marginal likelihood computation for model selection and hypothesis testing: an extensive review. *SIAM Review, 65*, 3–58.

Marin, J. M., Mengersen, K., & Robert, C. P. (2005). Bayesian modelling and inference on mixtures of distributions. In *Handbook of statistics* (Vol. 25, pp. 459–507). Elsevier.

Martin, G. M., Frazier, D. T., Maneesoonthorn, W., Loaiza-Maya, R., Huber, F., Koop, G., & Panagiotelis, A. (2024). Bayesian forecasting in economics and finance: A modern review. *International Journal of Forecasting, 40*, 811–839.

Maruotti, A., Punzo, A., & Bagnato, L. (2019). Hidden Markov and semi-Markov models with multivariate leptokurtic-normal components for robust modeling of daily returns series. *Journal of Financial Econometrics, 17*, 91–117.

McLachlan, G. J., & Peel, D. (2000). *Finite mixture models*. Wiley.

Metropolis, N., Rosenbluth, A. W., Rosenbluth, M. N., Teller, A. H., & Teller, E. (1953). Equation of state calculations by fast computing machines. *The Journal of Chemical Physics, 21*, 1087–1092.

Morgan, J. (1996). *Riskmetrics-technical document*. Morgan Guaranty Trust Company, Technical report.

Nguyen, T. M., & Wu, Q. J. (2013). Bounded asymmetrical Student's-t mixture model. *IEEE Transactions on Cybernetics, 44*, 857–869.

Peel, D., & McLachlan, G. J. (2000). Robust mixture modelling using the t distribution. *Statistics and Computing, 10*, 335–348.

Pennoni, F., Bartolucci, F., Forte, G., & Ametrano, F. (2022). Exploring the dependencies among main cryptocurrency log-returns: A hidden Markov model. *Economic Notes, 51*, Article e12193.

R Core Team. (2025). *R: A Language and Environment for Statistical Computing*. R Foundation for Statistical Computing.

Rodríguez, C. E., & Walker, S. G. (2014). Label switching in Bayesian mixture models: Deterministic relabeling strategies. *Journal of Computational and Graphical Statistics, 23*, 25–45.

Seara Vieira, A., Borrajo, L., & Iglesias, E. (2016). Improving the text classification using clustering and a novel HMM to reduce the dimensionality. *Computer Methods and Programs in Biomedicine, 136*, 119–130.

Sperrin, M., Jaki, T., & Wit, E. (2010). Probabilistic relabelling strategies for the label switching problem in Bayesian mixture models. *Statistics and Computing, 20*, 357–366.

Spiegelhalter, D. J., Best, N. G., Carlin, B. P., & Van Der Linde, A. (2002). Bayesian measures of model complexity and fit. *Journal of the Royal Statistical Society, Series B, 64*, 583–639.

Stephens, M. (2000a). Bayesian analysis of mixture models with an unknown number of components–an alternative to reversible jump methods. *Annals of Statistics, 28*, 40–74.

Stephens, M. (2000b). Dealing with label switching in mixture models. *Journal of the Royal Statistical Society, Series B, 62*, 795–809.

Tancini, D., Bartolucci, F., & Pandolfi, S. (2024). A comparison between marginal likelihood and data augmented MCMC algorithms for Gaussian hidden Markov models. *Journal of Statistical Computation and Simulation, 94*, 1571–1594.

Tanner, M. A., & Wong, W. H. (1987). The calculation of posterior distributions by data augmentation. *Journal of the American Statistical Association, 82*, 528–540.

Vats, D., Flegal, J. M., & Jones, G. L. (2019). Multivariate output analysis for Markov chain Monte Carlo. *Biometrika, 106*, 321–337.

Vehtari, A., Gabry, J., Magnusson, M., Yao, Y., Bürkner, P. C., Paananen, T., Gelman, A. (2015). loo: Efficient leave-one-out cross-validation and WAIC for Bayesian models. https://cran.r-project.org/package=loo. CRAN: Contributed Packages.

Vehtari, A., Gelman, A., & Gabry, J. (2017). Practical Bayesian model evaluation using leave-one-out cross-validation and WAIC. *Statistics and Computing, 27*, 1413–1432.

Watanabe, S. (2013). A widely applicable Bayesian information criterion. *Journal of Machine Learning Research, 14*, 867–897.

Watanabe, S., & Opper, M. (2010). Asymptotic equivalence of Bayes cross validation and widely applicable information criterion in singular learning theory. *Journal of Machine Learning Research, 11*, 3571–3594.

Yao, Y., Zhao, X., Wu, Y., Zhang, Y., & Rong, J. (2021). Clustering driver behavior using dynamic time warping and hidden Markov model. *Journal of Intelligent Transportation Systems, 25*, 249–262.

Young, S., & Woodland, P. (1994). State clustering in hidden Markov model-based continuous speech recognition. *Computer Speech and Language, 8*, 369–383.

Zeng, Y., & Garcia-Frias, J. (2006). A novel HMM-based clustering algorithm for the analysis of gene expression time-course data. *Computational Statistics & Data Analysis, 50*, 2472–2494.

Zhang, H., Wu, Q. J., Nguyen, T. M., & Sun, X. (2013). Synthetic aperture radar image segmentation by modified Student's t-mixture model. *IEEE Transactions on Geoscience and Remote Sensing, 52*, 4391–4403.

Zheng, Y., Jeon, B., Sun, L., Zhang, J., & Zhang, H. (2017). Student's t-hidden Markov model for unsupervised learning using localized feature selection. *IEEE Transactions on Circuits and Systems for Video Technology, 28*, 2586–2598.

Zucchini, W., Macdonald, I., & Langrock, R. (2016). *Hidden Markov models for time series: An introduction using R*. Chapman & Hall/CRC.

Link Prediction in Temporal Networks: A Dynamic Stochastic Block Model Approach

Silvia Pandolfi, Luca Brusa, Fulvia Pennoni, and Francesco Bartolucci

Abstract Longitudinal social networks observed over discrete time occasions can be effectively modeled using dynamic stochastic block models, where nodes are clustered into latent groups (or blocks) evolving via a hidden Markov process. We propose a novel link prediction method that forecasts future connections by estimating node memberships and their evolution over time. The approach relies on a variational approximation of the log-likelihood function for efficient maximum likelihood parameter estimation and predicts edges based on the most likely latent group assignments. Through simulations across various scenarios and an application to ant interaction data, our approach demonstrates accurate link predictions, particularly in networks with strong community structure and temporal persistence.

Keywords Complex networks · Hidden Markov chain · Maximum likelihood estimation · Predicted connections · Variational inference

Supplementary Information The online version contains supplementary material available at https://doi.org/10.1007/978-3-032-24141-2_6.

S. Pandolfi (✉) · F. Bartolucci
University of Perugia, Perugia, Italy
e-mail: silvia.pandolfi@unipg.it

F. Bartolucci
e-mail: francesco.bartolucci@unipg.it

L. Brusa · F. Pennoni
University of Milano-Bicocca, Milan, Italy
e-mail: luca.brusa@unimib.it

F. Pennoni
e-mail: fulvia.pennoni@unimib.it

F. Bartolucci et al. (eds.), *Models for Longitudinal Data with Applications to Early Warning Systems*, Springer Proceedings in Mathematics & Statistics 553,
https://doi.org/10.1007/978-3-032-24141-2_6

1 Introduction

Over the past 20 years, statistical analysis of networks has emerged as a fundamental framework for studying complex interaction systems across a wide range of disciplines. In this context, a network is represented as a set of nodes connected by edges (or links), where each node corresponds to a unit within the system and each edge represents a pairwise interaction between them. Since this definition is broad and flexible, network structures can model a variety of phenomena in various fields. In sociology, for example, nodes may represent individuals and edges indicate relationships like friendships or collaborations. In economics, networks can capture trade or financial connections, with nodes corresponding to firms, countries, or banks, and edges indicating transactions, ownership ties, or credit dependencies. Here, we propose a link prediction method for dynamic networks, also referred to as temporal or longitudinal networks, where a series of network snapshots, corresponding to different sets of connections, is observed at discrete time points. We focus in particular on simple graphs, namely, undirected binary graphs without self-loops or parallel edges, specifically based on the stochastic block model (SBM; Holland et al., 1983; Nowicki & Snijders, 2001; Snijders & Nowicki, 1997). This model may be formulated using discrete latent variables (Bartolucci et al., 2022), where nodes are clustered into unobservable groups (or blocks), and the probabilities of connections between pairs of nodes only depend on their cluster memberships. While our focus is on simple graphs, the SBM has been extended in various ways to accommodate more complex network structures, including weighted and dynamic networks, as well as multiplex and multipartite graphs, and hypergraphs; see Lee and Wilkinson (2019) and Brusa and Pennoni (2025) for a review of such extensions and some related applications. In this work, we focus on longitudinal data, where node connections may vary across successive discrete time points. In such a context, the dynamic SBM (DSBM; Bartolucci & Pandolfi, 2020; Matias & Miele, 2017; Yang et al., 2011) can account for time-varying block memberships, where each node can belong to distinct blocks across time, identified by node- and time-specific discrete latent variables following a Markov chain with initial and transition probabilities.

In network analysis, an important task is link prediction. In static networks, which are observed at a single point in time, the aim is to identify missing or unobserved connections between nodes, while in dynamic networks, whose structure evolves over time, link prediction aims at estimating the probability of future interactions, as well as the reemergence of previously dormant links. This is particularly valuable in early warning contexts, as it may help to detect high-risk or anomalous connections, while also providing uncertainty measures before traditional monitoring systems can react. Moreover, link prediction plays a significant role in network reconstruction (Guimerà & Sales-Pardo, 2009), where it is useful for detecting missing or spurious links, and in network completion (Kim & Leskovec, 2011), where it is applied to rebuild networks from partial or noisy data. For instance, Guimerà and Sales-Pardo (2009) applied these techniques to the protein interaction network of yeast to identify unreliable interactions and improve data quality. Other practical examples of the importance of forecasting include recommending friends or items on social platforms (Li et al., 2014), and detecting spam in email communications (Huang & Zeng, 2006).

In the following, we briefly describe some approaches adopted for link prediction that can also be useful for detecting critical changes considering first static and then dynamic networks. Comprehensive reviews of link prediction methods can be found in Lü and Zhou (2011), Kumar et al. (2020), and Wu et al. (2022). In static networks, similarity-based approaches are commonly used due to their simplicity and computational efficiency; see, among others, Zhou et al. (2009) and Liu and Lü (2010). These methods rely on scores that estimate the probability of connections between nodes, based on their structural proximity within the network. Such scores may be derived from various metrics, such as the number of shared neighbors, the overlap between their neighborhoods, or weighted versions of the connections. Other methods aim to reduce the network dimensionality to facilitate prediction, as proposed by, among others, Fukumizu et al. (2004) and Pecli et al. (2015). Supervised machine learning models are also used to forecast potential links by leveraging features derived from network topology and node attributes; see Hasan et al. (2006) and Doppa et al. (2010). On the other hand, model-based approaches provide a probabilistic framework to allocate nodes into unobservable clusters and exploit the resulting cluster structure; see, among others, Clauset et al. (2008) and Williamson (2016). Focusing specifically on SBMs, link prediction has been extensively studied by, among others, Guimerà and Sales-Pardo (2009), Zhao et al. (2017), Vallès et al. (2018), Tarrés-Deulofeu et al. (2019), and Nair et al. (2023). The models are typically estimated using techniques such as maximum likelihood or Bayesian inference, and the estimated model parameters are used to predict edge probabilities. Moreover, Nair et al. (2023) extended this approach by incorporating similarity indices as node covariates in the estimation of SBMs. A well-known drawback of model-based approaches is their computational complexity, which can make them impractical for very large networks. However, they offer valuable insights into the underlying structure of the network, that are generally inaccessible to simpler similarity-based algorithms.

In the dynamic setting, where the network is observed over a sequence of T time snapshots, link prediction refers to the task of predicting the probability of future connections at time $T + 1$. A major challenge in this context is that not only edges, but also nodes, may appear and disappear over time. Most of the existing literature, however, overlooks this aspect, assuming that the set of nodes remains the same at all time points (Divakaran & Mohan, 2020), which may lead to a loss of valuable information. In the following, we briefly describe the most common approaches, and we refer the reader to Divakaran and Mohan (2020) for a more comprehensive review. Matrix and tensor factorization techniques, such as singular value decomposition (Golub & Reinsch, 1970), are used in Dunlavy et al. (2011) to generate future link scores from the adjacency matrices corresponding to different time points. Topological measures and similarity indexes can also be applied with longitudinal data (Ibrahim & Chen, 2015; He et al., 2024). Among model-based methods, Safdari et al. (2022) presented a framework for directed longitudinal networks that employs a Markov process to model the appearance and disappearance of edges over time. In their approach, connection probabilities depend on the community memberships of nodes and the existence of the reciprocal edge at the previous time point. However, key limitations of this model include its restriction to directed networks and the assumption that community memberships remain constant over time.

In this work, we introduce a new approach for link prediction in longitudinal networks. The proposal is based on the DSBM, which is estimated from network data observed over T time points. Estimation is performed using an approximate maximum likelihood approach, relying on a variational expectation–maximization algorithm (Yang et al., 2011). The resulting parameter estimates are then used to predict connection probabilities at time $T + 1$, and an edge is predicted when this probability exceeds a certain threshold, which is determined based on the F1-score or similar metrics. The proposed method is validated through an extensive simulation study covering a wide range of settings, as well as an application to interactions among a colony of ants. The code employed for the simulations and for developing the application has been implemented for the software R (R Core Team, 2025).

The remainder of this chapter is organized as follows. Section 2 introduces the DSBM and the maximum likelihood estimation procedure. Section 3 presents the proposed link prediction approach. Section 4 shows the results of the simulation study and Sect. 5 illustrates the application. Section 6 provides some concluding remarks.

2 Dynamic Stochastic Block Model

In this section, we outline the main assumptions underlying DSBMs for longitudinal social network data. We then discuss issues related to model inference and illustrate the estimation approach based on a variational approximation.

2.1 Main Assumptions

Let $\boldsymbol{Y}^{(t)}$ denote the $n \times n$ adjacency matrix observed at occasion t, with $t = 1, \ldots, T$, where n is the number of nodes and T the number of time occasions. The generic element $y_{ij}^{(t)}$ of this matrix, where $i, j = 1, \ldots, n$ and $i \neq j$, is the realization of the binary response variable $Y_{ij}^{(t)}$ equal to 1 if there is an edge between nodes i and j at occasion t, and 0 otherwise. We focus on undirected relations without self-loops, so that $\boldsymbol{Y}^{(t)}$ is a symmetric matrix with missing values on the main diagonal, although the method we illustrate may be easily extended to deal with directed networks and to account for the presence of self-loops (Brusa & Pennoni, 2025). Finally, let $\mathcal{Y} = \{\boldsymbol{Y}^{(1)}, \ldots, \boldsymbol{Y}^{(T)}\}$ be the set of all network snapshots taken across time.

DSBMs (e.g., Bartolucci et al., 2018; Bartolucci and Pandolfi, 2020; Matias and Miele, 2017; Yang et al., 2011) assume that each node in the network belongs to one of k distinct blocks. Block memberships are represented by node- and time-specific discrete latent variables $U_i^{(t)} \in \{1, \ldots, k\}$, which are collected in the mutually independent and identically distributed vectors $\boldsymbol{U}_i = (U_i^{(1)}, \ldots, U_i^{(T)})'$. Each random vector $\boldsymbol{U}_i$ follows a Markov chain, with initial probabilities defined as

$$\pi_u = p(U_i^{(1)} = u), \quad u = 1, \ldots, k,$$

and transition probabilities denoted by

$$\pi_{v|u} = p(U_i^{(t)} = v \mid U_i^{(t-1)} = u), \quad t = 2, \ldots, T, \; u, v = 1, \ \ldots, k.$$

DSBMs also assume *conditional independence* between nodes: conditional on the latent variables, the response variables $Y_{ij}^{(t)}$ are independent and identically distributed. Moreover, given the latent blocks occupied by the two nodes, each of these variables follows a Bernoulli distribution, denoted by $\mathcal{B}(\cdot)$, with edge probability only depending on the hidden blocks (or states). Formally, we have:

$$Y_{ij}^{(t)} \mid U_i^{(t)} = u, U_j^{(t)} = v \sim \mathcal{B}(\beta_{uv}),$$

where $\beta_{uv} = p(Y_{ij}^{(t)} = 1 \mid U_i^{(t)} = u, U_j^{(t)} = v)$ are the edge probabilities. Note that here we consider binary dynamic networks with time-homogeneous model parameters and a fixed set of nodes that does not change over time. However, several generalizations have been proposed in the literature. For instance, weighted networks can be accommodated by specifying an alternative conditional distribution for the random variables $Y_{ij}^{(t)}$. Additionally, to enhance model flexibility, the constraint that parameters of the response distribution are time constant could be relaxed. An extra parameter can also be introduced to account for potential network sparsity; see, among others, Ambroise and Matias (2012), Xu and Hero (2014), Matias and Miele (2017) for a detailed description of these approaches.

2.2 Model Inference

Let $\mathcal{U} = \{\boldsymbol{U}_i, i = 1, \ldots, n\}$ denote the overall set of latent variables in the model; the observed network distribution, corresponding to the model likelihood, is obtained by marginalizing out these latent variables from the joint distribution of $\mathcal{Y}$ and $\mathcal{U}$, that is,

$$p(\mathcal{Y}) = \sum_{\mathcal{U}} p(\mathcal{Y}, \mathcal{U}), \quad p(\mathcal{Y}, \mathcal{U}) = p(\mathcal{Y} \mid \mathcal{U}) p(\mathcal{U}), \tag{1}$$

where $\sum_{\mathcal{U}}$ denotes the sum over the support of $\mathcal{U}$ and

$$p(\mathcal{Y} \mid \mathcal{U}) = \prod_{t=1}^{T} \prod_{i=1}^{n-1} \prod_{j=i+1}^{n} p(y_{ij}^{(t)} \mid u_i^{(t)}, u_j^{(t)}),$$

$$p(\mathcal{U}) = \prod_{i=1}^{n} \pi_{u_i^{(1)}} \prod_{t=2}^{T} \pi_{u_i^{(t)} | u_i^{(t-1)}}.$$

Evaluating the sum in Eq. (1), and then the likelihood function, becomes computationally prohibitive even for relatively small values of n, T, and k. Consequently, full maximum likelihood estimation is generally infeasible, except for networks of a very limited size.

2.2.1 Variational Approximation

In the context of parameter estimation for SBMs, a widely adopted approach involves the use of a variational approximation of the Expectation–Maximization (EM) algorithm (Dempster et al., 1977), commonly referred to as the Variational EM (VEM) algorithm; see Daudin et al. (2008) for the implementation in the context of static SBMs, and Yang et al. (2011) and Bartolucci and Pandolfi (2020) for the extensions to the dynamic setting. Indeed, due to the computational difficulties in obtaining the marginal network distribution, the posterior expectation of the complete data log-likelihood, which is required for the classical EM algorithm, is itself intractable. Variational inference overcomes this issue by optimizing a tractable lower bound of the log-likelihood computed under a suitable approximate posterior distribution.

Let $p(\mathcal{U} \mid \mathcal{Y})$ denote the posterior distribution of $\mathcal{U}$ given the observed data $\mathcal{Y}$ and let $\mathbb{Q}(\mathcal{U})$ denote its approximation. The VEM algorithm maximizes the following lower bound of the log-likelihood function:

$$\begin{aligned} \mathcal{J}(\boldsymbol{\theta}) &= \log p(\mathcal{Y}) - KL\left[\mathbb{Q}(\mathcal{U}) \mid\mid p(\mathcal{U} \mid \mathcal{Y})\right] \\ &= \log p(\mathcal{Y}) - \sum_{\mathcal{U}} \mathbb{Q}(\mathcal{U}) \log \frac{\mathbb{Q}(\mathcal{U})}{p(\mathcal{U} \mid \mathcal{Y})} \\ &= \sum_{\mathcal{U}} \mathbb{Q}(\mathcal{U}) \log p(\mathcal{Y}, \mathcal{U}) - \sum_{\mathcal{U}} \mathbb{Q}(\mathcal{U}) \log \mathbb{Q}(\mathcal{U}), \end{aligned} \tag{2}$$

where $\boldsymbol{\theta}$ denotes the vector of all free model parameters and $KL[\cdot \mid\mid \cdot]$ denotes the Kullback–Leibler divergence between two distributions. Regarding $\mathbb{Q}(\mathcal{U})$, the classical variational approach developed by Yang et al. (2011) is based on the mean-field approximation, that is, the class of approximate distributions that assumes a posteriori independence across individuals and time occasions:

$$\mathbb{Q}(\mathcal{U}) = \prod_{t=1}^{T}\prod_{i=1}^{n}\mathbb{Q}(U_i^{(t)}) = \prod_{t=1}^{T}\prod_{i=1}^{n}\prod_{u=1}^{k}\tau(t,i,u)^{I(U_i^{(t)}=u)}.$$

The variational parameter $\tau(t,i,u)$ can be interpreted as an approximation of $p(U_i^{(t)} = u \mid \mathcal{Y})$, which is intractable and therefore replaced by parameters that can be estimated. Obviously, these parameters satisfy the following constraints: $\tau(t,i,u) \geq 0$, for all i, t, and u, and $\sum_{u=1}^{k}\tau(t,i,u) = 1$, for all i and t. Using these quantities, Eq. (2) can be rewritten as

$$\begin{aligned}\mathcal{J}(\boldsymbol{\theta}) &= \sum_{i=1}^{n}\sum_{u=1}^{k}\tau(1,i,u)\log\pi_u \\ &+ \sum_{t=2}^{T}\sum_{i=1}^{n}\sum_{u=1}^{k}\sum_{v=1}^{k}\tau(t-1,i,u)\tau(t,i,v)\log\pi_{v|u} \\ &+ \sum_{t=1}^{T}\sum_{i=1}^{n-1}\sum_{j=i+1}^{n}\sum_{u=1}^{k}\sum_{v=1}^{k}\tau(t,i,u)\tau(t,j,v)\log\phi(y_{ij}^{(t)},\beta_{uv}) \\ &- \sum_{t=1}^{T}\sum_{i=1}^{n}\sum_{u=1}^{k}\tau(t,i,u)\log\tau(t,i,u),\end{aligned}$$

where $\phi(y,\beta_{uv}) = \beta_{uv}^{y}(1-\beta_{uv})^{1-y}$ corresponds to $p(Y_{ij}^{(t)} = y \mid U_i^{(t)} = u, U_j^{(t)} = v)$, with $y = 0, 1$.

To obtain parameter estimates, the VEM algorithm alternates two steps until convergence: the E-step and the M-step. The E-step consists in maximizing $\mathcal{J}(\boldsymbol{\theta})$ with respect to $\tau(t,i,u)$, with $\boldsymbol{\theta}$ fixed at the values obtained from the previous iteration. This leads to the following updating rule:

$$\tau(t,i,u) \propto \exp\left[\sum_{\substack{j=1\\ j\neq i}}^{n}\sum_{v=1}^{k}\tau(t,j,v)\log\phi(y_{ij}^{(t)},\beta_{uv}) + r_{iu}^{(t)}\right], \tag{3}$$

with

$$r_{iu}^{(t)} = \begin{cases}\log\pi_u + \sum_{v=1}^{k}\tau(2,i,v)\log\pi_{v|u}, & t=1,\\ \sum_{v=1}^{k}\tau(t-1,i,v)\log\pi_{u|v} + \sum_{v=1}^{k}\tau(t+1,i,v)\log\pi_{v|u}, & 1<t<T,\\ \sum_{v=1}^{k}\tau(T-1,i,v)\log\pi_{u|v}, & t=T.\end{cases}$$

In Eq. (3), the symbol $\propto$ indicates that the quantities $\tau(t,i,u)$ are obtained by normalizing the expression at the right-hand side.

At the M-step of the VEM algorithm, we maximize $\mathcal{J}(\boldsymbol{\theta})$ with respect to $\boldsymbol{\theta}$ with the $\tau(t, i, u)$ fixed at the value obtained from the E-step. The corresponding parameter estimates are updated as

$$\pi_u = \frac{\sum_{i=1}^{n} \tau(1, i, u)}{n}, \quad u = 1, \ldots, k,$$

$$\pi_{v|u} = \frac{\sum_{i=1}^{n} \sum_{t=2}^{T} \tau(t-1, i, u)\tau(t, i, v)}{\sum_{i=1}^{n} \sum_{t=2}^{T} \tau(t-1, i, u)}, \quad u, v = 1, \ldots, k,$$

$$\beta_{uv} = \frac{\sum_{i=1}^{n-1} \sum_{j=i+1}^{n} \sum_{t=1}^{T} \tau(t, i, u)\tau(t, j, v) y_{ij}^{(t)}}{\sum_{i=1}^{n-1} \sum_{j=i+1}^{n} \sum_{t=1}^{T} \tau(t, i, u)\tau(t, j, v)}, \quad u, v = 1, \ldots, k.$$

An alternative approximation for defining $\mathbb{Q}(\mathcal{U})$ was proposed by Matias and Miele (2017), based on the assumption of *a posteriori* independence across individuals, but not across time occasions. Under this assumption, the authors developed an algorithm to maximize the resulting lower bound of the log-likelihood function, employing approximate iterative steps. More recently, Bartolucci and Pandolfi (2020) introduced an exact algorithm for the maximization of this target function, thereby reducing the risk of getting a suboptimal solution.

In this work, since our main goal is link prediction, we rely on the VEM algorithm proposed by Yang et al. (2011), which, despite potentially yielding approximate parameter estimates, is well-suited to our purposes and significantly reduces the computational cost.

2.2.2 Algorithm Initialization

As is common in discrete latent variable models, the target function of the DSBM tends to be highly multimodal. Therefore, a multi-start initialization strategy of the VEM algorithm, which combines both deterministic and random rules, is necessary. Relying on the deterministic rule, nodes are first clustered using a k-means algorithm, and the model parameters are then estimated based on the resulting partition. In particular, the k-means algorithm is applied to the rows of an extended data matrix formed by row-wise concatenation of the adjacency matrices $\boldsymbol{Y}^{(t)}$ over time. This approach yields an initial clustering of the $n \times T$ observations, effectively capturing the dynamic structure of the data (Bartolucci et al., 2018). The random rule is instead based on generating initial solutions by assigning nodes in the network to one of the k blocks of the DSBM according to a time-varying random allocation. As in the deterministic case, initial parameter values are derived from this clustering, and the VEM algorithm is subsequently executed until convergence.

2.2.3 Model Selection and Clustering

It is important to note that, in the proposed approach for link prediction, the number of latent blocks of the DSBM is assumed to be known. However, there are many situations where this number is not known a priori. This issue has been widely discussed in the literature. In the context of variational inference, the Integrated Classification Likelihood (ICL) criterion (Biernacki et al., 2000) is typically used to select the optimal value of k. Other criteria may also be considered; however, especially when dealing with complex networks, model selection remains a challenging problem, which is beyond the scope of the present chapter.

A further important aspect related to estimation involves clustering of the observed nodes into the estimated blocks, which can be achieved through the proposed variational inference scheme. As is standard in the literature, local decoding is used to predict the latent block for each unit i and time occasion t, following the maximum a posteriori (MAP) rule. In particular, this rule relies on the estimated values of the multinomial probabilities $\tau(t, i, u)$, denoted by $\hat{\tau}(t, i, u)$: the predicted block at occasion t for node i, namely $\hat{u}_{it}$, is obtained as the value of u that maximizes $\hat{\tau}(t, i, u)$. Note that this approach implicitly assumes independence of the terms $\tau(t, i, u)$ across units and time occasions. A more general alternative is provided by global decoding, which jointly determines the most likely configuration of latent blocks over all units and times.

3 Link Prediction

Under the framework of temporally evolving networks with a latent community structure, an important task is link prediction. This involves estimating the probability that a new edge will appear between two nodes at time $T + 1$, given the observed sequence of network snapshots up to time T; see among others, Xu and Hero (2014).

3.1 Out-of-Sample Forecast

The aim is to predict, on the basis of the observed data, the link (edge) probability, $p_{ij}^{(T+1)}$, that is, the probability that $Y_{ij}^{(T+1)} = 1$ for a given couple of nodes (i, j). From a model-based perspective, we can exploit the maximum likelihood estimates of the DSBM obtained at convergence of the VEM algorithm to perform this task. More in detail, the parameters inferred from the model applied to the dynamic network up to time T can be used to perform (out-of-sample) prediction of the posterior block membership probabilities for the next time occasion as follows:

$$\hat{\tau}(T+1, i, u) = \sum_{\bar{u}=1}^{k} \hat{\tau}(T, i, \bar{u})\hat{\pi}_{u|\bar{u}}, \quad u = 1, \ldots, k,$$

where $\hat{\pi}_{u|\bar{u}}$ denotes the estimated transition probabilities of the Markov chain. This provides a predictive distribution over each node's block at time $T + 1$.

We propose to estimate the one-step-ahead probability of a potential edge between nodes i and j using two different approaches: conditional or unconditional on the predicted latent memberships. In the conditional case, the estimated probability of a link between nodes i and j at time $T + 1$ is given by

$$\hat{p}_{ij}^{(T+1)} = \hat{\beta}_{\hat{u}_i^{(T+1)}, \hat{u}_j^{(T+1)}},$$

where $\hat{\beta}_{uv}$ are the estimated edge probabilities and $\hat{u}_i^{(T+1)}$ is the predicted block for node i at occasion $T + 1$, computed on the basis of the MAP rule as $\hat{u}_i^{(T+1)} = \text{argmax}_u \ \hat{\tau}(T+1, i, u)$. The unconditional link prediction is instead computed as follows:

$$\tilde{p}_{ij}^{(T+1)} = \sum_{u=1}^{k} \sum_{v=1}^{k} \hat{\tau}(T+1, i, u)\hat{\tau}(T+1, j, v)\hat{\beta}_{uv}.$$

This approach yields a probabilistic prediction for each potential edge (i, j) at time $T + 1$, which can be used to determine the most likely connections at the subsequent time occasion. Obviously, this procedure may be extended to the general case of forecasting future observations at time $t = T + 1, \ldots, T^*$, with $T^* > T$. Moreover, we highlight that the proposed estimation and prediction procedures do not depend on the specific structure of the network (directed or undirected, with or without self-loops), and can therefore be straightforwardly extended to more general network settings.

Finally, it should be emphasized that the choice between the two approaches described above may depend on the specific task. When the goal is to predict the link probability for a specific node pair (i, j), or to produce a ranking of the node pairs likely to form future links, either the conditional or the unconditional prediction method may be employed. In such a situation, the corresponding performance may depend on the level of uncertainty in the posterior block allocation. In particular, when posterior memberships are highly uncertain, marginalizing over them, as done in the unconditional approach, could better capture the uncertainty in the underlying block structure. On the other hand, if the objective is to predict the joint probability of future connections over a subgraph or the entire graph, the conditional approach is more appropriate as the DSBM assumes that edge existence is conditionally independent across node pairs given the block assignments. Therefore, caution is advised when using the unconditional approach, as it may overlook the underlying dependencies captured by the block structure.

3.2 Evaluation Metrics for Temporal Link Prediction

In order to assess the performance of the proposed link prediction, we rely on standard evaluation metrics, commonly used to quantify the accuracy of prediction algorithms. In particular, we consider the *area under the ROC curve* (AUC) (Lü & Zhou, 2011), which is the most frequently used threshold-free metric in link prediction. This measure can be interpreted as the probability that a randomly chosen missing link (i.e., a node pair that will be connected in the future) is assigned a higher score than a randomly chosen non-existent link (i.e., a node pair that will remain unconnected), given the ranking of all the non-observed links. In addition to AUC, the F1-score (Davis & Goadrich, 2006) is another important measure of prediction accuracy. As is well known, it takes into account both *precision*, that is, the proportion of predicted links that actually exist in the network, and *recall*, that is, the proportion of true (future) links that the model correctly predicts. Obviously, to calculate both precision and recall, it is necessary to set a threshold $c \in [0, 1]$ that maps predicted probabilities $\hat{p}_{ij}^{(T+1)}$ (or $\tilde{p}_{ij}^{(T+1)}$) into a binary variable: the model predicts a future link between nodes i and j if $\hat{p}_{ij}^{(T+1)} > c$ (or $\tilde{p}_{ij}^{(T+1)} > c$); these measures, as well as the F1-score, can only be computed when the true future links are known.

Here, we propose to calibrate c by considering the in-sample link predictions for $t = 1, \ldots, T$. More in detail, we estimate the (in-sample) link probabilities $\hat{p}_{ij}^{(t)}$ or $\tilde{p}_{ij}^{(t)}$, for $t = 1, \ldots, T$, using both the conditional and the unconditional approach defined previously. Then, we calibrate the classification threshold c in order to obtain the optimal value, denoted by c^*, which maximizes the F1-score based on the comparison between the (in-sample) predictions and the observed links. For further details on the choice of the optimal threshold c^*, see also Chap. Exploring Binary Regression and Hidden Markov Models for Early Warning Systems of this book.

4 Simulation Study

In order to evaluate the performance of the proposed link prediction approach, we carried out a simulation study that considers different network sizes, numbers of latent blocks, and underlying block structures. The generated network snapshots are partitioned into a training set consisting of the first $T - 1$ time occasions, used to calibrate the threshold c, and a test set corresponding to the last observed time T. In such a way, we evaluate the proposed approach in predicting the existence of a connection between node pairs (i, j) at a future time occasion, using the AUC and F1-score, and other performance metrics.

4.1 Simulation Design

We randomly generated $B = 100$ networks of size $n = 100, 200$ nodes from a DSBM with $T = 10, 20$ time steps. At the initial time occasion, nodes were assigned to latent blocks uniformly at random, i.e., $\pi_u = 1/k$, for $u = 1, \ldots, k$, with a number of blocks set to $k = 3$ or $k = 4$. We also considered two distinctive levels of group stability, corresponding to highly and weakly persistent blocks, respectively. In particular, the assumed transition matrix has elements parameterized through an index ρ as follows:

$$\pi_{v|u} = \frac{\rho^{|v-u|}}{\sum_{\bar{v}} \rho^{|\bar{v}-u|}}, \quad u, v = 1, \ldots, k,$$

so that high values of ρ correspond to a low probability of persistence of the series, and vice versa. In this study, we examined two different values of ρ, equal to 0.1 and 0.3. Finally, we considered three different community structures by varying the intra-group and inter-group connectivity parameters, leading to the following model specifications:

- model M_1: defined by setting $\beta_{uv} = 0.65 \times \alpha$ if $u = v$, and $\beta_{uv} = 0.05$ if $u \neq v$, with $\alpha \sim Unif(1, 1.5)$. This model exhibits strong intra-group connectivity and weak inter-group connectivity;
- model M_2: defined by setting $\beta_{uv} = 0.05 \times \alpha$ if $u = v$, and $\beta_{uv} = 0.65$ if $u \neq v$, resulting in high inter-group connectivity and weak intra-group connectivity;
- model M_3: characterized by a predefined, heterogeneous community structure. In particular, high intra-group connection probabilities (0.97) ensure a strong community structure. Moreover, the inter-group connection probabilities are set to introduce varying levels of separation: moderate connectivity (≈ 0.82), weak connectivity (≈ 0.34), and near disconnection (≈ 0.05) between different pairs of blocks.

Accordingly, the results reported in the following are based on a series of scenarios:

- the first scenario, which serves as a benchmark, considers $n = 100$ nodes, $T = 10$ time snapshots, $k = 3$ latent blocks, and high persistence probabilities, with $\rho = 0.1$. The connectivity parameters correspond to model M_1, characterized by a well-defined community structure with high intra-group and low inter-group connection probabilities;
- the second scenario evaluates the performance of the proposed approach as the sample size increases, setting $n = 200$ nodes, while keeping the other parameters the same as in the first scenario;
- the third scenario examines the effect of increasing the number of hidden blocks to $k = 4$, keeping $n = 100$ and $T = 10$ as in the benchmark;
- the fourth scenario differs from the benchmark in terms of the number of time occasions, setting $T = 20$;

- the fifth scenario assesses the impact of decreased persistence probabilities by setting $\rho = 0.3$;
- the sixth and seventh scenarios evaluate the effect of different community structures by assuming models M_2 and M_3, respectively.

4.2 Simulation Results

To evaluate the performance of the proposed link prediction approach, we analyzed the predicted connection probabilities between node pairs in the test set, corresponding to the last time occasion, using AUC and F1-score metrics, computed for both conditional and unconditional predictions. Results are reported in Table 1.

We also report the F1-score values obtained for the in-sample predictions on the first $T - 1$ time occasions, which were used to calibrate the threshold c^*. Additionally, we provide the average value of this optimal threshold selected in the simulated samples. Finally, we report further evaluation metrics, including sensitivity (recall), specificity, precision, and accuracy, referred to the out-of-sample forecasting.

The results indicate that both the conditional and unconditional methods perform well in predicting future links, consistently achieving F1-score and AUC values above 0.6. The only exception is the fifth scenario, which assumes low persistence probabilities in the Markov chain. Under this setting, the value of the F1-score drops significantly, especially when relying on predicted probabilities computed unconditionally on the predicted block memberships. The AUC metric also records the lowest value compared to the other scenarios. Moreover, we observe that the F1-score improves with an increasing number of nodes and, in particular, when the community structure corresponds to model M_3. Overall, the conditional approach outperforms the unconditional one in terms of F1-score, whereas this latter method yields better results when evaluated using the AUC metric.

5 Application

We present an application concerning data collected from daily interactions within a colony of $n = 164$ ants. The experiment was conducted under controlled laboratory conditions at the University of Lausanne[1] (Mersch et al., 2013). Ants constitute a complex biological system that has been extensively studied, particularly because their colonies form dynamic networks generated by patterns of individual encounters and interactions (Frank & Kronauer, 2024; Gordon, 2010). In the dataset under consideration, the position and orientation of each ant were repeatedly recorded over

[1] Data are freely available at the following link: https://datadryad.org/dataset/doi:10.5061/dryad.8d8h7#citations.

Table 1 Performance metrics for link prediction, on the basis of predicted probabilities computed conditionally ($\hat{p}_{ij}^{T+1}$) and unconditionally ($\tilde{p}_{ij}^{T+1}$) on the predicted block memberships, under different simulated scenarios. The benchmark scenario corresponds to $n = 100$ nodes, $T = 10$ time occasions, $k = 3$ latent blocks, $\rho = 0.1$, and model M_1

	Benchmark		$n = 200$		$k = 4$	
	$\hat{p}_{ij}^{T+1}$	$\tilde{p}_{ij}^{T+1}$	$\hat{p}_{ij}^{T+1}$	$\tilde{p}_{ij}^{T+1}$	$\hat{p}_{ij}^{T+1}$	$\tilde{p}_{ij}^{T+1}$
F1-score (in-sample)	0.850	0.850	0.850	0.850	0.834	0.834
F1-score (out-of-sample)	0.674	0.669	0.679	0.679	0.651	0.641
c^* (average)	0.500	0.500	0.500	0.500	0.500	0.500
AUC	0.780	0.803	0.787	0.807	0.783	0.821
Sensitivity/Recall	0.708	0.696	0.711	0.711	0.661	0.639
Specificity	0.829	0.833	0.832	0.832	0.881	0.888
Precision	0.644	0.647	0.649	0.649	0.641	0.647
Accuracy	0.792	0.791	0.795	0.795	0.828	0.827
	$T = 20$		$\rho = 0.3$		M_2	
	$\hat{p}_{ij}^{T+1}$	$\tilde{p}_{ij}^{T+1}$	$\hat{p}_{ij}^{T+1}$	$\tilde{p}_{ij}^{T+1}$	$\hat{p}_{ij}^{T+1}$	$\tilde{p}_{ij}^{T+1}$
F1-score (in-sample)	0.847	0.847	0.825	0.825	0.773	0.773
F1-score (out-of-sample)	0.676	0.673	0.473	0.275	0.695	0.695
c^* (average)	0.496	0.496	0.451	0.451	0.500	0.500
AUC	0.783	0.803	0.622	0.642	0.672	0.692
Sensitivity/Recall	0.712	0.698	0.523	0.154	0.856	0.856
Specificity	0.826	0.824	0.699	0.744	0.494	0.494
Precision	0.645	0.642	0.438	0.440	0.585	0.585
Accuracy	0.791	0.785	0.646	0.564	0.659	0.659
	M_3					
	$\hat{p}_{ij}^{T+1}$	$\tilde{p}_{ij}^{T+1}$				
F1-score (in-sample)	0.883	0.883				
F1-score (out-of-sample)	0.805	0.805				
c^* (average)	0.500	0.500				
AUC	0.794	0.799				
Sensitivity/Recall	0.782	0.782				
Specificity	0.760	0.760				
Precision	0.829	0.829				
Accuracy	0.774	0.774				

a period of 41 days. An interaction is defined as occurring whenever the front end of one ant enters the proximity region of another. Ants primarily communicate through indirect interactions involving chemical substances known as pheromones, which are produced and secreted by specialized glands (Kirman, 1993; Morgan, 2009). For each day, a binary network $\boldsymbol{Y}^{(t)}$ is built such that an edge between individuals i and j is included if at least one contact was recorded between the corresponding ants on day t. Additional observed information is also available, including ant age and size, number of visits to the brood, nest entrance, and rubbish piles, as well as number of foraging events. The number of ants decreases after the first 11 days, likely due to mortality. For this reason, in the analyses that follow, we consider only the data from these 11 days. To predict connections, we first estimate the DSBM using data from the first 10 days ($T = 10$) and present the results along with a discussion of the estimated parameters. We then show the prediction results for day 11, obtained adopting the approach described in Sect. 3.

5.1 Results: Model Estimation

The DSBM is estimated considering $k = 4$ latent blocks; this choice for the number of components is motivated by previous studies, which highlight a clear division of labor within the ant colony, with distinct groups of ants performing specific tasks. After estimating the model, nodes (ants) are dynamically assigned to the four groups based on posterior probabilities through the MAP rule (see Sect. 2). The estimated parameters are shown in Fig. 1, including connection probabilities $\hat{\beta}_{uv}$ (panel a),

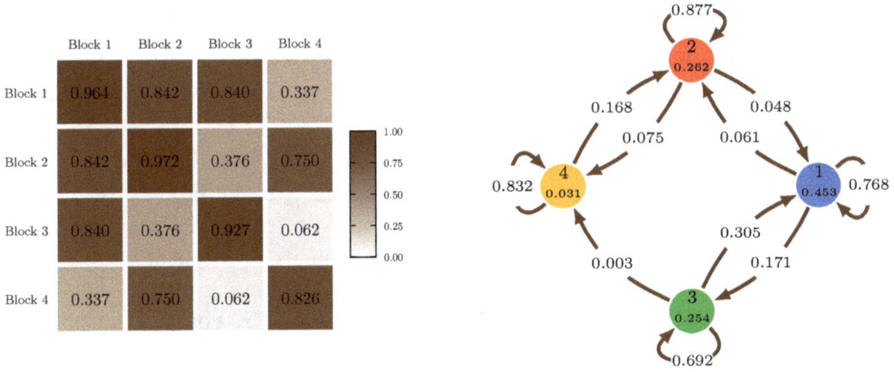

(a) Estimated edge probabilities ($\hat{\beta}_{uv}$): darker shades indicate higher probabilities

(b) Estimated initial ($\hat{\pi}_u$, within the nodes) and transition ($\hat{\pi}_{v|u}$, along the edges) probabilities

Fig. 1 Summary of the estimated model parameters for the ant colony data under the DSBM with $k = 4$ latent blocks

Table 2 Average values within each latent block for ant age and size, number of foraging events, and number of visits to the brood, nest entrance, and rubbish piles

	Latent block			
	1	2	3	4
Age (days)	92.66	153.10	86.62	220.80
Body size (mm)	10.42	10.39	10.46	10.66
Foraging events	3.11	34.48	0.02	29.00
Brood visits	172.68	4.82	475.00	0.26
Nest entrance visits	4.69	35.40	0.18	8.27
Rubbish piles visits	16.50	4.35	0.62	1.64

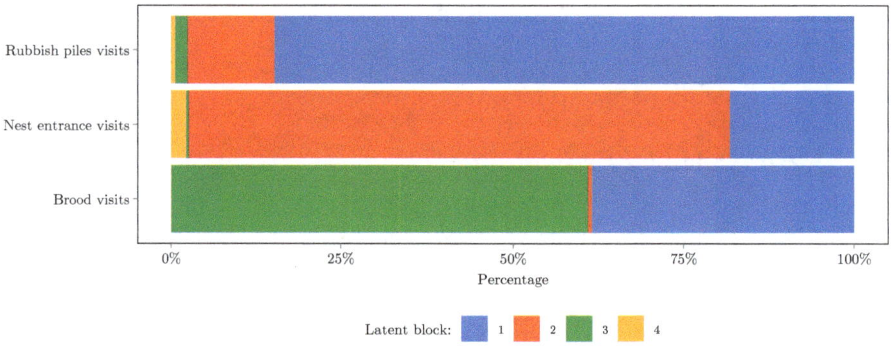

Fig. 2 Percentage of visits to different nest areas (brood, nest entrance, and rubbish piles) made by ants in each latent block

initial probabilities $\hat{\pi}_u$ (panel *b*, within the nodes), and transition probabilities $\hat{\pi}_{v|u}$ (panel *b*, along the edges). Additionally, Table 2 and Fig. 2 summarize supplementary information about the ants within each latent block, providing further insight into their characteristics.

In the following, the main characteristics of the blocks are described. According to Fig. 2, the 3rd latent block presents the highest average number of visits to the brood (475.00), with ants in this group accounting for approximately 60% of all such visits. In contrast, the average number of visits to the nest entrance and rubbish piles is almost negligible, as is the number of foraging events. Therefore, this group predominantly consists of nurses, characterized by younger ants with an average age of 86.62 days. The internal connection probability is very high (0.927), and there are also strong interactions with the 1st block (0.840). The 1st group, in turn, is responsible for approximately the remaining 40% of visits to the brood, corresponding to a daily average of 172.68. However, ants in this group are primarily characterized by the highest number of visits to the rubbish piles, with an average of 16.50, accounting for about 85% of all such visits. These ants are therefore mainly cleaners, with an average age of approximately 92.66 days. Connection probabilities are very high both within the group and with blocks 2 and 3. The remaining two blocks are characterized by a high number of foraging events, with daily averages of 34.48 and 29.00 for groups 2 and 4, respectively. However, they differ significantly in the number of visits to the nest entrance: ants in group 2 account for an average of 35.40 visits (approximately 79% of the total), compared to just 8.27 visits (about 2% of the total) for those in group 4. Additionally, ants in group 4 are considerably older, with an average age of 220.80 days, compared to an average of 153.10 days for those in group 2. Ants in these two blocks are predominantly foragers. Those in group 2 tend to forage near the nest, frequently returning to it, while those in group 4 undertake longer, less frequent trips, accumulating more resources and returning to the nest less often. The 4th latent block thus exhibits high connection probabilities only internally (0.826) and with group 2 (0.750). Block 2, in turn, shows strong connection probabilities with all blocks except the 3rd.

The estimated averaged transition matrix is depicted in Fig. 1 (panel *b*). All blocks exhibit high temporal persistence. The highest transition probabilities occur between the 1st and 3rd latent blocks, reflecting transitions from nurses to cleaners (with a probability of 0.305) and the reverse transition from cleaners to nurses (with a probability of 0.171). Moderate transitions are also observed between the 2nd and 4th groups.

5.2 Results: Link Prediction

In this section, we illustrate the use of the estimated model parameters from the previous section to forecast edges for day $T + 1 = 11$. Following the approach described in Sect. 3, we compute both versions of the prediction probabilities, $\hat{p}_{ij}^{(T+1)}$ and $\tilde{p}_{ij}^{(T+1)}$, that is, conditionally and unconditionally on the predicted latent membership at time $T + 1$. A connection between (i, j) is then predicted when the corresponding probability exceeds a specified threshold, which is determined based on the F1-score evaluated on the basis of the in-sample link prediction.

Table 3 summarizes the main evaluation metrics obtained with this procedure.

Table 3 Main performance metrics for prediction of connections among ants during the 11th day for the conditional and unconditional case

	$\hat{p}_{ij}^{(T+1)}$	$\tilde{p}_{ij}^{(T+1)}$
Sensitivity/Recall	0.930	0.998
Specificity	0.439	0.056
Precision	0.794	0.711
Accuracy	0.782	0.715
F1-score	0.856	0.831
AUC	0.761	0.825

Both methods yield high and comparable F1-scores, 0.856 and 0.831, respectively. The number of correctly predicted links is particularly high, as reflected in the sensitivity values: using the conditional approach, approximately 93.0% of ant connections are correctly predicted, increasing to 99.8% with the unconditional method. This latter procedure shows notably lower specificity, indicating a higher number of false positives, that is, predicted connections that do not actually occur. The overall accuracy (i.e., the proportion of correctly predicted links and non-links) is relatively high in both cases: 0.782 and 0.715, respectively. Overall, the first approach shows slightly better and more balanced performance, despite predicting a slightly lower number of total connections.

6 Conclusions

In this chapter, we consider network analysis as an integral part of statistical modeling for examining interactions between statistical units. Specifically, using a Dynamic Stochastic Block Model (DSBM), we focus on link prediction, a key task in temporal network analysis. Our goal is to estimate the probability of future interactions, as well as the potential reemergence of previously inactive links. This capability is particularly valuable in early warning scenarios, as it enables the detection of high-risk or anomalous connections and provides measures of uncertainty before conventional monitoring systems can respond.

The present work proposes two prediction methods for temporal networks: one conditional and the other unconditional on the predicted latent memberships. When the goal is to estimate the link probability for a specific node pair or to rank node pairs by their chance of forming future links, both the conditional and the unconditional approaches can be applied. However, if the aim is to estimate the joint probability of future connections across a subgraph or the entire graph, the conditional approach is more appropriate, due to the underlying assumptions of the DSBM.

We show a good predictive accuracy in both approaches through simulations on synthetic data, under varying network characteristics. Furthermore, we illustrate the performance of the proposed methods in an application involving ant interaction networks. The interaction patterns of ants have been shown to resemble those observed in other complex systems—including neural networks in the brain, human social networks, and even internet traffic (Gordon, 2010).

Future research may consider the inclusion of network-specific information, such as similarity measures between nodes, or relevant node-level attributes in the model formulation. This additional structure could enhance interpretability and predictive performance, and may be particularly useful in case of time-heterogeneous model parameters. Another possible direction involves adapting the proposed approach to better account for sparsity, which could lead to improved estimation in settings where the network is sparse. Other relevant research directions concern the selection of the optimal threshold c^*, which is currently based on in-sample predictive performance. More advanced strategies, such as cross-validation procedures or cost-sensitive approaches, which assign different weights to false positives and false negatives, could provide a more robust criterion for threshold selection. Finally, the proposed link prediction procedure may be extended to the general case of forecasting future edges at time $t = T + 1, \ldots, T^\star$, with $T^\star > T$, with suitable strategies to mitigate error propagation across the prediction horizon.

References

Ambroise, C., & Matias, C. (2012). New consistent and asymptotically normal parameter estimates for random-graph mixture models. *Journal of the Royal Statistical Society: Series B, 74*, 3–35.

Bartolucci, F., Marino, M. F., & Pandolfi, S. (2018). Dealing with reciprocity in dynamic stochastic block models. *Computational Statistics & Data Analysis, 123*, 86–100.

Bartolucci, F., & Pandolfi, S. (2020). An exact algorithm for time-dependent variational inference for the dynamic stochastic block model. *Pattern Recognition Letters, 138*, 362–369.

Bartolucci, F., Pandolfi, S., & Pennoni, F. (2022). Discrete latent variable models. *Annual Review of Statistics and Its Application, 9*, 425–452.

Biernacki, C., Celeux, G., & Govaert, G. (2000). Assessing a mixture model for clustering with the integrated completed likelihood. *IEEE Transactions on Pattern Analysis and Machine Intelligence, 22*, 719–725.

Brusa, L., & Pennoni, F. (2025). Stochastic block model based on variational inference and its extensions: An application to examine global migration dynamics. In M. Nakai (Ed.), *Advances in quantitative approaches to sociological issues. Behaviormetrics: Quantitative approaches to human behavior* (Vol. 20, pp. 1–27). Springer: Singapore.

Clauset, A., Moore, C., & Newman, M. E. J. (2008). Hierarchical structure and the prediction of missing links in networks. *Nature, 453*, 98–101.

Daudin, J.-J., Picard, F., & Robin, S. (2008). A mixture model for random graphs. *Statistics and Computing, 18*, 173–183.

Davis, J., & Goadrich, M. (2006). The relationship between precision-recall and ROC curves. In *Proceedings of the 23rd International Conference on Machine Learning* (pp. 233–240).

Dempster, A. P., Laird, N. M., & Rubin, D. B. (1977). Maximum likelihood from incomplete data via the EM algorithm. *Journal of the Royal Statistical Society: Series B, 39*, 1–38.

Divakaran, A., & Mohan, A. (2020). Temporal link prediction: A survey. *New Generation Computing, 38*, 213–258.

Doppa, J. R., Yu, J., Tadepalli, P., & Getoor, L. (2010). Learning algorithms for link prediction based on chance constraints. In J. L. Balcázar, F. Bonchi, A. Gionis, & M. Sebag (Eds.), *Proceedings of the 2010 European Conference on Machine Learning and Knowledge Discovery in Databases* (pp. 344–360). Berlin: Springer.

Dunlavy, D. M., Kolda, T. G., & Acar, E. (2011). Temporal link prediction using matrix and tensor factorizations. *ACM Transactions on Knowledge Discovery from Data, 5*, 1–27.

Frank, D. D., & Kronauer, D. J. C. (2024). The budding neuroscience of ant social behavior. *Annual Review of Neuroscience, 47*, 167–185.

Fukumizu, K., Bach, F. R., & Jordan, M. (2004). Dimensionality reduction for supervised learning with reproducing kernel Hilbert spaces. *Journal of Machine Learning Research, 5*, 73–99.

Golub, G. H., & Reinsch, C. (1970). Singular value decomposition and least squares solutions. *Numerische Mathematik, 14*, 403–420.

Gordon, D. M. (2010). *Ant encounters: Interaction networks and colony behavior*. Princeton, New Jersey: Princeton University Press.

Guimerà, R., & Sales-Pardo, M. (2009). Missing and spurious interactions and the reconstruction of complex networks. *Proceedings of the National Academy of Sciences, 106*, 22073–22078.

Hasan, M. A., Chaoji, V., Salem, S., & Zaki, M. (2006). Link prediction using supervised learning. In *Proceedings of the 4th Workshop on Link Analysis, Counterterrorism and Security.*

He, X., Ghasemian, A., Lee, E., Clauset, A., & Mucha, P. J. (2024). Sequential stacking link prediction algorithms for temporal networks. *Nature Communications, 15*, 1–15.

Holland, P., Laskey, K. B., & Leinhardt, S. (1983). Stochastic blockmodels: First steps. *Social Networks, 5*, 109–137.

Huang, Z., & Zeng, D. D. (2006). A link prediction approach to anomalous email detection. In *2006 IEEE International Conference on Systems, Man and Cybernetics* (pp. 1131–1136). Institute of Electrical and Electronics Engineers.

Ibrahim, N. M., & Chen, L. (2015). Link prediction in dynamic social networks by integrating different types of information. *Applied Intelligence, 42*, 738–750.

Kim, M., & Leskovec, J. (2011). The network completion problem: Inferring missing nodes and edges in networks. In B. Liu, H. Liu, C. Clifton, T. Washio, & C. Kamath (Eds.), *Proceedings of the 2011 SIAM International Conference on Data Mining* (pp. 47–58). Society for Industrial and Applied Mathematics.

Kirman, A. (1993). Ants, rationality, and recruitment. *The Quarterly Journal of Economics, 108*, 137–156.

Kumar, A., Singh, S. S., Singh, K., & Biswas, B. (2020). Link prediction techniques, applications, and performance: A survey. *Physica A: Statistical Mechanics and its Applications, 553*, 1–46.

Lee, C., & Wilkinson, D. J. (2019). A review of stochastic block models and extensions for graph clustering. *Applied Network Science, 4*, 1–50.

Li, J., Zhang, L., Meng, F., & Li, F. (2014). Recommendation algorithm based on link prediction and domain knowledge in retail transactions. *Procedia Computer Science, 31*, 875–881.

Liu, W., & Lü, L. (2010). Link prediction based on local random walk. *Europhysics Letters, 89*, 1–6.

Lü, L., & Zhou, T. (2011). Link prediction in complex networks: A survey. *Physica A: Statistical Mechanics and its Applications, 390*, 1150–1170.

Matias, C., & Miele, V. (2017). Statistical clustering of temporal networks through a dynamic stochastic block model. *Journal of the Royal Statistical Society: Series B, 79*, 1119–1141.

Mersch, D. P., Crespi, A., & Keller, L. (2013). Tracking individuals shows spatial fidelity is a key regulator of ant social organization. *Science, 340*, 1090–1093.

Morgan, E. D. (2009). Trail pheromones of ants. *Physiological Entomology, 34*, 1–17.

Nair, L. S., Jayaraman, S., & Krishna Nagam, S. P. (2023). An improved link prediction approach for directed complex networks using stochastic block modeling. *Big Data and Cognitive Computing, 7*, 1–18.

Nowicki, K., & Snijders, T. A. B. (2001). Estimation and prediction for stochastic blockstructures. *Journal of the American Statistical Association, 96*, 1077–1087.

Pecli, A., Giovanini, B., Pacheco, C., Moreira, C., Ferreira, F., Tosta, F., Tesolin, J., Dias, M., Filho, S., Cavalcanti, M. C., & Goldschmidt, R. (2015). Dimensionality reduction for supervised learning in link prediction problems. In S. Hammoudi, L. Maciaszek, & E. Teniente (Eds.), *ICEIS 2015–17th International Conference on Enterprise Information Systems, Proceedings* (pp. 295–302). Science and Technology Publications.

R Core Team. (2025). *R: A language and environment for statistical computing*. Vienna, Austria: R Foundation for Statistical Computing.

Safdari, H., Contisciani, M., & De Bacco, C. (2022). Reciprocity, community detection, and link prediction in dynamic networks. *Journal of Physics: Complexity, 3*, 1–10.

Snijders, T. A. B., & Nowicki, K. (1997). Estimation and prediction for stochastic blockmodels for graphs with latent block structure. *Journal of Classification, 14*, 75–100.

Tarrés-Deulofeu, M., Godoy-Lorite, A., Guimerà, R., & Sales-Pardo, M. (2019). Tensorial and bipartite block models for link prediction in layered networks and temporal networks. *Physical Review E, 99*, 1–11.

Vallès, T., Peixoto, T. P., Sales-Pardo, M., & Guimerà, R. (2018). Consistencies and inconsistencies between model selection and link prediction in networks. *Physical Review E, 97*, 1–12.

Williamson, S. A. (2016). Nonparametric network models for link prediction. *The Journal of Machine Learning Research, 17*, 7102–7121.
Wu, H., Song, C., Ge, Y., & Ge, T. (2022). Link prediction on complex networks: An experimental survey. *Data Science and Engineering, 7*, 253–278.
Xu, K. S., & Hero, A. O. (2014). Dynamic stochastic blockmodels for time-evolving social networks. *IEEE Journal of Selected Topics in Signal Processing, 8*, 552–562.
Yang, T., Chi, Y., Zhu, S., Gong, Y., & Jin, R. (2011). Detecting communities and their evolutions in dynamic social networks - a Bayesian approach. *Machine Learning, 82*, 157–189.
Zhao, Y., Wu, Y.-J., Levina, E., & Zhu, J. (2017). Link prediction for partially observed networks. *Journal of Computational and Graphical Statistics, 26*, 725–733.
Zhou, T., Lü, L., & Zhang, Y. C. (2009). Predicting missing links via local information. *The European Physical Journal B, 71*, 623–630.

The Substitution Between Primary and Emergency Care in Individuals with Chronic Conditions: Evidence from a Structural Model

Mauro Laudicella and Paolo Li Donni

Abstract We examine the substitution between primary and emergency secondary care in patients with chronic conditions by using a dynamic structural model for longitudinal count data. The hidden Markov bivariate Poisson model allows for disentangling unobservable time-varying heterogeneity from the dynamic effect of utilization of primary and secondary care. Also, unobserved heterogeneity can be modeled as a function of observable individual characteristics, including age, income, and education. We find that an additional standard visit in primary care reduces utilization of emergency care by 9%, while the substitution effect of preventive assessment visits is two times larger. The substitution effect is larger in individuals with a high utilization profile, it decreases with individual age, while it shows little difference by socioeconomic status. Finally, the probability of switching utilization profile decreases with age. Our analysis contributes to the existing evidence on the substitution effect from policy evaluation studies.

Keywords Primary care · Secondary care · Chronic conditions · Hidden Markov models · Panel data

1 Introduction

The rising prevalence of chronic conditions in an aging population is considered a key driver of the growing demand for healthcare services in many high-income countries and a source of concern for the financial stability of publicly funded health systems (Kollerup et al., 2022; OECD, 2015). A widespread belief among policy-makers is that an effective disease management in primary care (PC) can reduce utilization of

M. Laudicella
Universidad Autónoma de Madrid, Madrid, Spain
e-mail: mauro.laudicella@uam.es

P. Li Donni (✉)
University of Palermo, Palermo, Italy
e-mail: paolo.lidonni@unipa.it

F. Bartolucci et al. (eds.), *Models for Longitudinal Data with Applications to Early Warning Systems*, Springer Proceedings in Mathematics & Statistics 553,
https://doi.org/10.1007/978-3-032-24141-2_7

more expensive emergency secondary care (ESC) in patients with such conditions (Dusheiko et al., 2011). For instance, a regular monitoring of the disease, encouraging behavioral changes, timely diagnosis of comorbidities, and referral to elective care, can reduce the risk of acute exacerbations leading to presentations in the hospital emergency department (ED) and emergency admissions (Reynolds et al., 2018). In these cases, using ESC is inefficient as it leads to higher costs for the taxpayer and suboptimal health outcomes for the patient as compared to disease management in PC.

The first contribution of this paper is study the substitution between PC and ESC in patients with chronic conditions by using a dynamic structural model of the demand of PC and ESC (Laudicella et al., 2024; Laudicella & LiDonni, 2022). The Hidden Markov Bivariate Poisson model (HMBPo) extends the class of finite mixture model for longitudinal count data to the bivariate case addressing some of the distinctive characteristics of the data-generating process highlighted in the literature. First, the demand of PC and ESC is jointly determined at the individual level, thus allowing for individuals accessing both types of care in the same time occasion (Atella & Dep, 2008). Second, state dependence within and between the demand of PC and ESC is captured by including a lag of PC and ESC in the outcome equations reflecting the fact that past utilization, namely, the past individual state, is often a strong predictor of present (Contoyannis et al., 2004; Hyppolite & Trivedi, 2012; Kohn & Liu, 2013). Third, unmeasured time-varying heterogeneity is captured by a discrete latent variable following a first-order Markov chain; hence, individuals are allowed to switch their latent state if they change their (unmeasured) health and preferences over time (Wouterse et al., 2013). Finally, time-specific non-permanent shocks that may affect both the demand of PC and ESC are captured by a residual correlation parameter. The HMBPo model allows us to extend the scope of existing studies by examining heterogeneity in the substitution effect at the level of patient and PC service delivered during the visit. We provide new evidence on the substitution effect in patients with different age and socioeconomic status, and with respect to standard PC visits, home visits, and preventive assessment visits. The second contribution of this paper is to extend the model in Laudicella and Li Donni (2022) to allow for the latent state transition probabilities to depend on individual characteristics. This new feature allows us to study heterogeneity in the time trajectories of latent utilization profiles and examine the probability that individuals switch to a high utilization profile as they become older.

We study a population of individuals with a clinical diagnosis of diabetes, chronic obstructive pulmonary disease (COPD), hypertension, and osteoporosis accessing healthcare services provided by the Danish National Health System (NHS). We follow these individuals from an initial health shock directly related to their condition to 5 years later, examining their utilization of PC and ESC over the period 2012–2017. The four chronic conditions examined in this study have a high prevalence in the population of many high-income countries and are often subject to special monitoring in PC and have been targeted by health policies aiming at reducing utilization of unnecessary ESC (Dusheiko et al., 2011; Tsiachristas et al., 2013). The Danish NHS offers an excellent setting to conduct our study as it provides a universal coverage to all residents in Denmark and is free of charge at the point of access, thus reducing confounding effects from differences in health insurance coverage and ability to pay for healthcare. Denmark also collects high-quality data on individual access to PC and ESC linked with individual socioeconomic characteristics, such as income and education. The Danish NHS and the Danish population share similar characteristics with many other high-income countries, such as the Nordic countries to whom they are closely related, and countries with similar universal healthcare system such as England, France, Italy, and Spain.

We find a noticeable substitution effect of PC on the demand for ESC with an additional PC visit reducing ESC utilization by 9.5% in individuals with chronic conditions. Preventive assessment visits in PC are two times more effective in reducing utilization of ESC than standard visits or home visits. The substitution effect decreases with age, while it shows no significant differences by income and education. Finally, we find evidence that individuals with chronic conditions are likely to change their utilization profile before reaching the senior age, with many of the switches occurring between 40 and 60 s and few in older ages. Our evidence contributes to the academic debate on the substitutability between PC and ESC and may inform future health policies targeting sociodemographic groups and health services with the largest scope for reducing utilization of ESC.

This chapter is organized as follows: Sect. 2 reviews the existing literature, Sect. 3 introduces the methods, Sect. 4 describes the data, Sect. 5 describes the empirical findings, while Sect. 6 discusses the main results and concludes. The full set of estimation functions has been implemented in a series of Mata routines for StataCorp. v.2018 available online.

2 Existing Literature

Numerous policy interventions have tried to reduce the demand for emergency care by improving access and quality of PC services for patients with chronic conditions, including the introduction of monetary incentives for monitoring their health (Dusheiko et al., 2011), additional fees for providing key services (Iezzi et al., 2014), risk-adjusted capitation fees (Pulleyblank et al., 2020; Rudkjøbing et al., 2015; Tummalapalli et al., 2022), and pay-for-coordination schemes promoting integration between primary and secondary care (Tsiachristas et al., 2013). Finally, policy interventions for patients with chronic diseases have been accompanied by additional measures directed to increasing accessibility of PC for the whole population, such as extending PC opening hours and opening new walking-in clinics (Baier et al., 2019; Dolton & Pathania, 2016; Lippi Bruni et al., 2016; Pinchbeck, 2019).

Most of the evidence on the scope for substituting ESC with PC comes from empirical studies that evaluate the effects of such policies after their implementation (Dolton & Pathania, 2016; Dusheiko et al., 2011; Fortney et al., 2005; Iezzi et al., 2014; Pinchbeck, 2019). The exogenous variation in the supply of PC introduced by the policy allows the researcher to identify the impact of utilization of PC on the demand for ESC, namely, the average treatment effect on the treated. This identification strategy addresses the confounding effect of unmeasured heterogeneity in patients' health driving the demand for PC and ESC in the same direction, thus resulting in a positive correlation and a spurious complementarity effect between the two variables. However, this strategy poses a number of limitations to the analysis of the substitution effect between PC and ESC. First, the timing of the policy and the specific health conditions and geographical areas targeted are rarely random, rather they are selected by the policy-maker to maximize the potential impact of the policy. Hence, the average substitution effect on the treated population is likely to be different from its potential effect on the whole population. Second, the substitution effect produced by the introduction of a policy incentive is tightly linked to the response of PC physicians to the incentive and the effective design of the policy. In other words, a small substitution effect could be evidence of a weak response to the incentive or a poor design of the policy, rather than a narrow scope for reducing ESC by empowering PC (Dusheiko et al., 2011; Gravelle et al., 2010; Rudkjøbing et al., 2015). Finally, the policy is usually designed to target PC physicians rather than patients, hence most of the existing empirical studies focus on aggregated data at the level of PC practice (Dolton & Pathania, 2016; Fortney et al., 2005; Iezzi et al., 2014; Whittaker et al., 2016). However, the demand of PC and ESC is the outcomes of decisions made by individual patients, rather than PC physicians. This discrepancy may introduce ecological fallacy in drawing conclusions about the substitution effect at the individual level and in assessing the heterogeneity of such an effect, for instance, in individuals with older ages and low socioeconomic status who are frequent users of both PC and ESC services.

The present study contributes to the existing literature by producing new evidence on the substitution effect based on a dynamic structural model, thus extending the scope of the analysis outside the policy evaluation framework adopted by existing studies.

3 Methods

Modeling the interdependence between the demand of PC and ESC over time requires some considerations on the data-generating process and the potential sources of correlations between these two outcome variables.

First, present utilization of PC and ESC is correlated with past utilization as highlighted in the literature (Contoyannis et al., 2004; Hyppolite & Trivedi, 2012; Kohn & Liu, 2013). For instance, a visit to the PC physician in the past month may prevent a visit to the hospital ED in the present month, and it may also require a second visit to the former if the health issue that prompted the first visit is not completely resolved. This source of state dependence *within* and *between* outcomes can be used to explain utilization-patterns between different types of care over time and the potential substitution effect between PC and ESC.

Second, observations are likely to be correlated over time reflecting systematic time-varying heterogeneity due to unmeasured health and preferences for utilization of care. This second source of correlation is likely to confound the correct identification of the potential substitution effect between PC and ESC. For instance, individuals with a strong health endowment may prefer to see their PC physician infrequently and bear no consequences on their risk of an ESC visit. In contrast, individuals with a weak health endowment who see their PC physician infrequently may end up in the hospital ED. Moreover, unmeasured health and preferences may be subject to systematic changes over time as the individual grows older and her chronic condition stabilizes or deteriorates, and such changes may differ according to observable individual characteristics. For instance, older individuals and individuals from a lower socioeconomic status may follow a different trajectory in their utilization profiles than other individuals. The literature models this source of heterogeneity by using *univariate* latent class models, which assume individuals are drawn from a discrete number of latent groups sharing similar unmeasured characteristics (e.g., Bago d'Uva (2005); Bago d'Uva and Jones (2009); Alfò & Maruotti, 2010; Hyppolite & Trivedi, 2012; Wouterse et al. (2013)).

Finally, a third source of correlation can be generated by unmeasured *time-specific* shocks that may affect utilization of both PC and ESC in a specific time period. For instance, individuals may experience an occasional health deterioration with short lasting consequences for their utilization of care.

Laudicella and Li Donni (2022) propose a Hidden Markov Bivariate Poisson (HMBPo) model to address the three sources of correlation descried above. The HMBPo model relies on two main assumptions: (i) unmeasured heterogeneity can be modeled by using a discrete distribution of latent states ξ_c collected in the vector $\boldsymbol{U}$, with $c = 1, \ldots, k$ following a first-order Markov chain and (ii) the latent states make PC and ESC conditionally independent given the full set of observable explanatory variables, which also include utilization of healthcare in past periods.

Suppose we observe two counts for every individual i, denoted by y_{jit}, with $j = 1, 2$, for each time occasion t, with $t = 1, \ldots, T_i$. In particular, y_{1it} and y_{2it} measure the individual's number of PC and ESC visits, respectively. Let denote with $\boldsymbol{y}_i$ the corresponding $T_i \times 2$ matrix defined by the following subvectors $(\boldsymbol{y}_{i1}, \boldsymbol{y}_{i2}, \ldots, \boldsymbol{y}_{iT_i})$. Let also indicate with $\boldsymbol{\alpha}_{it}$ a vector of time-varying random effects collecting α_{1it} and α_{2it} and with $\boldsymbol{x}_{jit}$ a vector of exogenous covariates of dimension $1 \times h_j$. The HMBPo can be described by two sets of equations modeling the relationship between unmeasured time-varying heterogeneity α_{1it} and α_{2it}, and outcomes y_1 and y_2.

The first set of equations describes the marginal distribution of y_{1it} and y_{2it} in *each* time period. In particular, this is assumed to follow a Bivariate Poisson (BPo), such that the probability of observing a specific realization of y_{1it} and y_{2it} is given by

$$p(\mathbf{y}_{it} \mid \boldsymbol{\alpha}_{it}, \mathbf{x}_{it}, \mathbf{y}_{i,t-1}) = \frac{\mu_{1it}^{y_{1it}} \mu_{2it}^{y_{2it}} e^{-\mu_{1it}-\mu_{2it}}}{y_{1it}!\, y_{2it}!} \times \left[1 + \lambda\left(e^{-y_{1it}} - e^{-q\mu_{1it}}\right)\left(e^{-y_{2it}} - e^{-q\mu_{2it}}\right)\right], \tag{1}$$

with $q = 1 - e^{-1}$ and $\mu_{jit} = \exp(\sum_{c=1}^{k} \alpha_{jit}(\xi_c) d_{jit}(\xi_c) + \mathbf{x}_{jit}\boldsymbol{\beta}_j + \gamma_{j1} y_{1i,t-1} + \gamma_{j2} y_{2i,t-1})$, with $j = 1, 2$ and $d_{jit}(\xi_c)$ denotes a dummy variable defining whether the ith unit belongs to the latent state ξ_c of the Markov chain at time t.

In this framework, the lag of the outcome variables, i.e., y_{1it-1} and y_{2it-1}, models state dependence within and between outcomes, $y_{\cdot it}$, hence the parameters γs capture the first source of correlation described above.

The individual random effects α_{1it} and α_{2it} capture the second source of correlation due to unmeasured time-varying heterogeneity; αs vary with the latent state ξ_c, hence they capture the persistence in using both PC and ESC that is due to individual unmeasured characteristics.

The second set of equations describes the distribution of $\boldsymbol{\alpha}$. Individuals are assumed to be drawn from a discrete latent distribution with k classes in each time period. These classes represent different unobserved states in which individuals can be observed with probability $p(\boldsymbol{\alpha}_{it} = \boldsymbol{\xi}_c)$ in state c. The latent states are assumed to follow a first-order Markov chain, such that

$$p(\boldsymbol{\alpha}_i = \boldsymbol{\xi}) = p^{(1)}(\boldsymbol{\alpha}_{i1} = \boldsymbol{\xi}_c) \prod_{t=2}^{T_i} p^{(t)}(\boldsymbol{\alpha}_{it} = \boldsymbol{\xi}_{\overline{c}} | \boldsymbol{\alpha}_{i,t-1} = \boldsymbol{\xi}_c). \tag{2}$$

This is described by two components: the initial and the transition probabilities. The initial probability simply refers to the probability of observing an individual in a specific state c in the first period. When covariates are included, this is usually parameterized by using a multinomial logit as a link function:

$$p^{(1)}(\boldsymbol{\alpha}_{i1} = \boldsymbol{\xi}_c | \mathbf{z}_{i1}, \mathbf{y}_{i0}) = \frac{\exp(\alpha_c + \boldsymbol{\beta}_c \mathbf{z}_{i1} + \boldsymbol{\gamma}_c \mathbf{y}_{i0})}{1 + \sum_{c=2}^{k} \exp(\alpha_c + \boldsymbol{\beta}_c \mathbf{z}_{i1} + \boldsymbol{\gamma}_c \mathbf{y}_{i0})}, \tag{3}$$

where $\mathbf{z}_{i1}$ is a vector of observable covariates. Estimating the coefficients from the initial condition and including the effect of $\mathbf{y}_{i0}$ in the unobserved process in the first period allows for the well-known initial condition problem (Heckman, 1981).

Finally, the transition probabilities model the unobserved heterogeneity over time. The original HMBPo model assumes that the probability to switch from state c to state $\overline{c}$ is homogenous, e.g., the same transition probability applies to each individual and time period (Laudicella & LiDonni, 2022). However, individuals with different sociodemographic characteristics may experience different time trajectories in their latent state, for instance, some age and socioeconomic groups may be more likely to change state than others. Examining heterogeneity in the trajectories of the latent states may be useful to identify sociodemographic groups with the highest probability

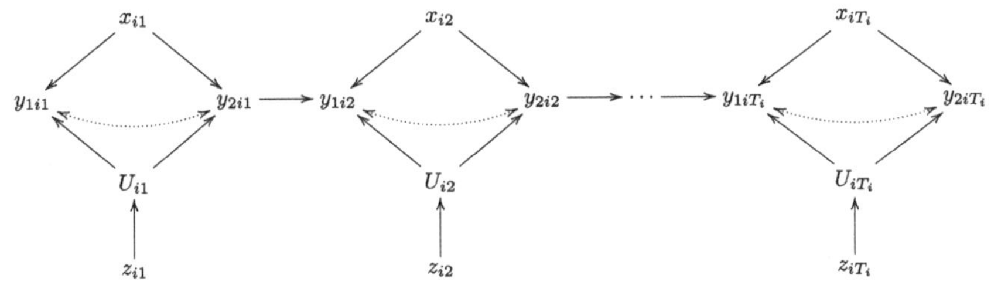

Fig. 1 Path diagram for the HMBPo model with individual transition probabilities

of changing their latent utilization profiles, and thus to inform policies aiming at reducing inequalities in access to care. To this end, we extend the original HMBPo model and allow transition probabilities to differ according to observable individual characteristics, such as age, income, and education. Hence, we move to a more general case with individual-specific transition matrix Π_{it}, which collects the following $k \times (k-1)$ probabilities:

$$p^{(t)}\left(\alpha_{it} = \xi_{\bar{c}}|\alpha_{i,t-1} = \xi_c, z_{it}\right) = \frac{\exp(\alpha_{c\bar{c}} + \beta_{c\bar{c}} z_{it})}{1 + \sum_{\bar{c}}^{k} \exp(\alpha_{c\bar{c}} + \beta_{c\bar{c}} z_{it})}, \tag{4}$$

with $c, \bar{c} = 1, \ldots k, c \neq \bar{c}$ and $t = 2, \ldots, T_i$. Transition probabilities can be collected on the matrix $\boldsymbol{\Pi}_i$. Notice that z is a vector of observable covariates influencing the transition from c to state $\bar{c}$; the homogeneity assumption is equivalent to setting $\beta_{c\bar{c}} = 0$ in Eq. (4).

Equations (1), (3), and (4) define the HMBPo model with individual transition probabilities. The structure of the model can be illustrated by the path diagram in Fig. 1. The arrowed lines from x_{it} to y_{ji1} represent the effects measured by the parameters on PC and ESC in Equation (1), while the dotted arrowed arcs depict the effects of time-specific shocks measured by λ. The initial condition in Eq. (3) is described by the arrow from z_{i1} to the latent variable U_{i1} collecting the latent states in the first period, while the subsequent arrows from z_{i2} to $U_{i2}, \ldots, z_{iT_i}$ to U_{iT_i} describe the effect of covariates on the probability of a switch between latent states in each time period (see Eq. (4)). The individual log-likelihood of the model can be described by the following equation:

$$\ell_i(\boldsymbol{\theta}) = \log\left\{\sum_{\alpha_{i1}} \cdots \sum_{\alpha_{iT_i}} \left[p^{(1)}(\alpha_{i1} = \xi_c|z_{i1}, y_{i0}) \prod_{t=2}^{T_i} p^{(t)}(\alpha_{it} = \xi_{\bar{c}}|\alpha_{i,t-1} = \xi_c, z_{it}) \times \right.\right.$$
$$\left.\left. \times \prod_{t=1}^{T_i} p(\boldsymbol{y}_{it}|\alpha_{it}, \boldsymbol{x}_{it}, \boldsymbol{y}_{i,t-1})\right]\right\},$$

where $\boldsymbol{\theta}$ collects all the non-redundant model parameters in $\boldsymbol{\alpha}$, $\boldsymbol{\beta}$, $\boldsymbol{\gamma}$, and λ. Estimate $\boldsymbol{\theta}$ requires to compute the above densities and probabilities in ℓ_i, which involves a sum over all the possible configurations of latent states over time. This computation can be efficiently achieved by using the Baum–Welch recursions (Welch, 2003); details are reported in Appendix 7. Then, the estimation is achieved by using the EM algorithm, which is a convenient algorithm maximizing a log-likelihood containing unobservable variables (Dempster et al., 1977). In our case, the unobservable components are represented by the latent states ξ, describing the distribution of the individual-specific parameters $\boldsymbol{\alpha}_{it}$.

We followed the algorithm provided in Laudicella and Li Donni (2022) to perform model estimation, while we followed Bartolucci and Farcomeni (2015) to check model identifiability and to compute the observed information matrix that provides standard errors for the parameter estimates. Specifically, we used Oakes' identity (Oakes, 1999), which is given by two components. The first component is the second derivative of the conditional expected value of the complete data log-likelihood given the observed data. This component is directly provided by the EM algorithm in Laudicella and Li Donni (2022). The second component is the first derivative of the score for the same expected log-likelihood with respect to the current value of the parameters.

The computation of marginal and transition probabilities is described in Appendix 7.

4 Data

We use data extracted from the Danish National Patient Register including PC visits and ESC visits. The data include information on visit date, hospital admission and discharge date, primary diagnosis and up to 20 secondary diagnoses reported using ICD-10 codes. All residents of Denmark are identified by a unique identification number that is used to follow them through the Danish National Patient Register and to link them to other registers at the individual level. We used linked data on individual annual income from the Income Statistics Register, level of education from the Education Register, and living alone status from the Central Person Register (for an overview on Danish registries see Thygesen et al., 2011). The characteristics of the Danish NHS and institutional framework are described in Appendix 8.

4.1 Study Population

Our study population includes individuals aged 40 and over with a clinical diagnosis of diabetes, COPD, hypertension, and osteoporosis, who experience an initial health shock directly related to their chronic condition in 2012. We follow individuals from the initial health shock in 2012 to 5 years later and examine their utilization of PC and

ESC over the period 2012–2017; the longitudinal analysis considers time intervals of 3 months. The chronic diseases included in our basket have a high prevalence in many high-income countries, they can be effectively managed in PC, and have a significant risk of acute exacerbations that may lead to an ESC visit. Such conditions have been targeted by policy-makers aiming at improving the quality of disease management in PC and reducing utilization of ESC (Dusheiko et al., 2011; Tsiachristas et al., 2013). Finally, our basket covers a wide spectrum of distinct conditions that do not overlap in their clinical definition and coding, while some patients can be affected by multiple chronic conditions at the same time.

We focus on individuals who experience a similar health shock in the same calendar year (2012) to allow for a precise measurement of their health status at the beginning of our study, including clinical diagnoses for all chronic conditions and comorbidities that are likely to affect individuals' patterns of utilization of PC and ESC. We define the initial health shock as the first ESC visit occurring in 2012 with a primary diagnosis for a chronic condition in our basket. Other studies have used a similar approach based on initial health shock to define their study population (Doyle, 2011; García-Gómez et al., 2013; Laudicella et al., 2022).

4.2 *Outcome and Control Variables*

We model PC visits and ESC visits as outcome variables, while lagged PC visits and lagged ESC visits are used as independent variables to capture state dependence within and between outcomes. PC visits includes standard face-to-face consultations at the PC practice, preventive assessment visits, and home visits and ESC visits include visits to the hospital ED for any cause including visits resulting in a hospital admission and visits treated at the hospital ED not resulting in a hospital admission. We study PC visits and ESC visits occurring after the initial health shock, from 2012 to 2017.

We use a basket of indicators capturing individual health and sociodemographic characteristics to control for observable individual heterogeneity in the analysis. Health indicators include selected chronic conditions (diabetes, COPD, hypertension, osteoporosis); total number of secondary diagnoses; indicators for highly prevalent acute conditions such as acute myocardial infarctions (AMI), strokes, and cancer; and Charlson index measuring individual mortality risk. The health indicators are measured at the time of the initial health shock and then updated over the study period (2012–2017). Demographic and socioeconomic indicators include age, gender, individual income, level of education, and living alone status. Individuals living alone are less likely to receive informal care that may influence their demand of PC and ESC (De Meijer et al., 2011). Socioeconomic indicators are measured at the time of the initial health shock and remain quite stable over the study period. Finally, we include an indicator of travel time to the nearest hospital ED to capture accessibility of ESC.

4.3 Descriptive Statistics

Table 1 reports descriptive statistics for our study population; individuals are observed in time periods of 3 months, from 2012 to 2017. On average, individuals experience 0.30 ESC visits and 2.25 PC visits per period; the latter include 2.09 standard face-to-face consultations at the PC practice, 0.12 home visits, and 0.09 preventive assessment visits. In terms of main demographic and health characteristics, 55% of individuals are female, the average age is 65.8 years old, 37% of individuals have diabetes, 28% COPD, 24% hypertension, and 17% osteoporosis, individuals have an average of 0.27 secondary diagnoses, 5% of individuals experienced an AMI, 10% a stroke, and 5% cancer. Moving to socioeconomic characteristics, individual average income is about 24,500 Euros, 44% of individuals completed the compulsory level of education only, while 40% completed upper secondary education, and 16% achieved a tertiary level of education. In Denmark, compulsory education includes primary and lower secondary school from age 7 to age 16; secondary education includes programs leading to general and vocational qualifications from age 17 to 19; and

Table 1 Summary statistics

Variable name	Mean	SD
ESC visits	0.30	0.81
PC visits	2.25	2.51
Standard visits	2.09	2.43
Home visits	0.12	0.54
Assessment visit	0.09	0.32
Female	0.55	–
Age	65.83	11.98
Living alone	0.46	0.50
Income	24,580.92	41,938.20
Compulsory education	0.44	0.50
Upper secondary education	0.40	0.49
Tertiary education	0.16	0.36
Travel time to ESC	13.41	10.28
Diabetes	0.37	–
COPD	0.28	–
Hypertension	0.24	–
Osteoporosis	0.17	–
Secondary diagnoses	0.27	0.97
Charlson index	1.22	1.28
AMI	0.05	–
Stroke	0.10	–
Cancer	0.05	–

Number of observations = 466,486; Number of patients = 31,182

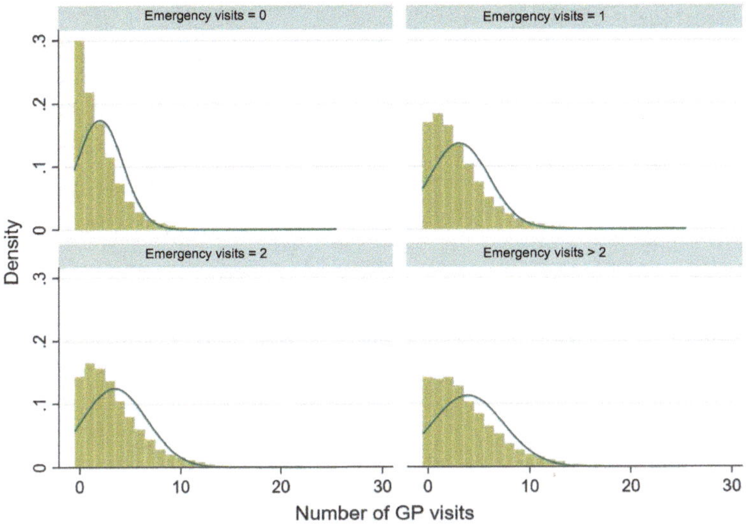

Fig. 2 Observed distribution of PC visits by increasing number of ESC visits

tertiary education includes short, medium, and long education programs oriented toward certain types of jobs, and including bachelor, master, and PhD degrees.

Figure 2 illustrates the distribution of PC visits conditional to the number of ESC visits; average and dispersion of the PC distribution increase with increasing number of ESC visits highlighting a positive correlation between these two outcomes as reported by other authors (Riphahn et al., 2003).

5 Results

We examined the relationship between PC and ESC by using the HMBPo model and compared its performance against two separate Poisson models with random effects (REPo). The latter is frequently used in the literature for modeling longitudinal count data on healthcare use, including PC and ESC visits, and thus it provides a good starting point for assessing the predictions of the HMBPo. The main difference between the two models is that the REPo model assumes independence between the demand of PC and ESC after conditioning on covariates, while the HMBPo allows for the interdependence between the two outcomes.[1]

[1] To control for the initial conditions in the REPo model, we follow the strategy proposed by Wooldridge (2005), which consists in including initial condition controls y_{1i0}, y_{2i0} and average of time-varying x among regressors for each equation of PC and ESC.

Table 2 Model selection criteria

Model	k	♯	Log-lik.	BIC	AIC
HMBPo	1	79	−1209026.74	2419084.66	2418211.47
	2	104	−1135760.58	2272878.68	2271729.17
	3	131	−1114281.77	2230273.49	2228825.55
	4	160	−1114258.90	2230606.28	2228837.80

Table 2 reports the BIC and the AIC for the HMBPo model with increasing number of latent states k. The BIC and AIC suggest that three latent states are sufficient to model the time-varying individual heterogeneity in the HMBPo model.

5.1 Determinants of Healthcare Utilization

Table 3 reports the estimated coefficients for state dependence, substitution effects, and socioeconomic variables in our study population of individuals with chronic conditions. Estimated coefficients can be interpreted as semi-elasticities.

Both the REPo and the HMBPo predict a positive state dependence in the demand for PC and ESC. An additional PC visit in the past period (namely, the past 3 months) results in a 4.3% increment in utilization of PC in the present period according to the HMBPo model, and a similar 4.5% increment according to the REPo. In contrast, an additional ESC visit in the past results in a 18% increment in utilization of ESC in the present according to the HMBPo and a 5% increment according to the REPo. The difference in the magnitude of state dependence for ESC is likely to reflect that HMBPo estimates are conditional to the PC distribution, while REPo is unconditional with a large unobserved heterogeneity, $\sigma^2 = 0.33$ for ESC. A positive state dependence in healthcare utilization is reported by other studies (Contoyannis et al., 2004; Hyppolite & Trivedi, 2012; Kohn & Liu, 2013).

Remarkable differences emerge in the substitution effects estimated by the two models. With respect to the effect of PC on ESC, the REPo predicts a small *complementarity effect* with an additional PC visit in the past 3 months resulting in a 2.4% increment in utilization of ESC in the present. In contrast, the HMBPo model predicts a *substitution effect* with an additional PC visit in the past 3 months reducing utilization of ESC by 9.5% in the present. A similar difference emerges with respect to the effect of ESC on PC: while the REPo predicts a small complementarity effect of 1.2%, the HMBPo model predicts a substitution effect of 5.2%.

The estimated 9.5% reduction in ESC associated with one additional PC visit should be interpreted as a structural causal semi-elasticity within our HMBPo framework, not as a policy-induced causal effect. However, our specification reduces key confounding channels by (i) conditioning on past utilization (state dependence), (ii) modeling time-varying unobserved heterogeneity via latent states and individual-

Table 3 Estimated coefficients

	REPo model		HMBPo model	
	PC visit	ESC visit	PC visit	ESC visit
ESC	0.0123***	0.0492***	–0.0522***	0.1821***
	(0.00)	(0.00)	(0.00)	(0.00)
PC	0.0452***	0.0241***	0.0434***	–0.0949***
	(0.00)	(0.00)	(0.00)	(0.00)
Female	0.0701***	–0.0140	0.0355***	–0.0352***
	(0.01)	(0.01)	(0.01)	(0.01)
Age	0.0425***	–0.0228***	0.0314***	–0.0261***
	(0.00)	(0.00)	(0.00)	(0.00)
Age^2	–0.0003***	0.0002***	–0.0002***	0.0002***
	(0.00)	(0.00)	(0.00)	(0.00)
Living alone	–0.0910***	0.1298***	–0.0187***	0.1956***
	(0.01)	(0.01)	(0.01)	(0.01)
Inc. 2nd quintile	0.0058	–0.0263	0.0066	–0.0456***
	(0.01)	(0.02)	(0.01)	(0.01)
Inc. 3rd quintile	0.0117	–0.0493***	0.0094	–0.0643***
	(0.01)	(0.02)	(0.01)	(0.01)
Inc. 4th quintile	–0.0162	–0.0574***	–0.0029	–0.0445***
	(0.01)	(0.02)	(0.01)	(0.01)
Inc. 5th quintile	–0.1240***	–0.2236***	–0.1192***	–0.2798***
	(0.01)	(0.02)	(0.01)	(0.01)
Upper secondary education	–0.0308***	–0.0311**	–0.0106**	0.0136*
	(0.01)	(0.01)	(0.01)	(0.01)
Tertiary education	–0.0862***	–0.0177	–0.0524***	0.0022
	(0.01)	(0.02)	(0.01)	(0.01)
Travel time to ESC	0.0058***	–0.0061***	0.0048***	–0.0067***
	(0.00)	(0.00)	(0.00)	(0.00)
Copd	0.0311***	0.3551***	0.0725***	0.3949***
	(0.01)	(0.01)	(0.01)	(0.01)
Hypertension	–0.0589***	–0.0591***	0.0014	0.0045
	(0.01)	(0.02)	(0.01)	(0.01)
Osteoporosis	–0.0341***	0.0832***	0.0425***	0.1924***
	(0.01)	(0.02)	(0.01)	(0.01)
Secondary diagnoses	0.0598***	0.3927***	0.0532***	0.3893***
	(0.00)	(0.00)	(0.00)	(0.00)
Charlson index	–0.0062**	0.1237***	0.0328***	0.1434***
	(0.00)	(0.01)	(0.00)	(0.00)
AMI	0.0971***	0.2782***	0.0106	0.0249*
	(0.01)	(0.03)	(0.01)	(0.01)
Stroke	–0.0054	0.1027***	–0.0212***	0.0484***
	(0.01)	(0.02)	(0.01)	(0.01)
Cancer	0.0268**	0.0849***	–0.0379***	–0.1161***
	(0.01)	(0.03)	(0.01)	(0.02)
Constant†	-1.3082***	-1.5928***	–0.6465***	–0.9258***
	(0.09)	(0.15)	(0.06)	(0.10)
σ^2	0.1026***	0.3383***		
	(0.01)	(0.01)		
λ				–0.0505*
				(0.03)

Standard errors are in parentheses. † average of the support points based on the posterior probabilities in the HMBPo. *significant at 10%, **significant at 5%, ***significant at 1%

specific transition probabilities, and (iii) allowing for residual, time-specific shocks common to PC and ESC. These features are consistent with a substitution mechanism and explain why the HMBPo reverses the complementarity suggested by simpler REPo models.

Moving to the sociodemographic variables, the REPo and the HMBPo models return similar estimated coefficients with similar direction and magnitude. The demand of PC and ESC decreases with age with diminishing marginal increments under both models. Individuals in the top quintile of the income distribution use 11.9% less PC and 27.9% less ESC than individuals in the bottom quintile (HMBPo model). Individuals with tertiary education use 5.24% less PC than individuals with compulsory education (baseline), while small and non-significant differences are found in utilization of ESC. In Denmark, access to PC and ESC is free of charge at the point of use, hence socioeconomic differences in utilization of care may capture heterogeneity in unmeasured health and preferences (Fooken & Jeet, 2022). Similar evidence of lower utilization of PC and ESC in higher socioeconomic groups are found in the literature (Bago d'Uva & Jones, 2009; Walsh & Laudicella, 2017). Individuals living alone use 1.8% less PC and 19.5% more ESC (HMBPo model); this variable is likely to capture the impact of having less access to informal care from a family member, which may be particularly useful to individuals with a chronic condition (De Meijer et al., 2011). Finally, individuals who have to drive an extra 15 min to reach the closest hospital ED use 0.4% more PC and 0.6 less ESC (HMBPo model).

Moving to the residual correlation between PC and ESC, the parameter λ in the HMBPo model shows a small magnitude and a 10% significance level. This suggests that the residual correlation between PC and ESC that is due to non-permanent time-specific shocks is small after modeling the other sources of correlation including state dependence, substitution effects, and time-varying latent states. Hence in our study population, the correlation between PC and ESC is sufficiently captured by these other factors in the HMBPo model.

Table 4 reports the results of our heterogeneity analysis of the substitution effect. First, we unpack the PC variable and examine different types of PC visits, including standard face-to-face visits, home visits, and preventive assessment visits (Columns 1, Table 4). A standard visit reduces utilization of ESC by 10%, which is similar to the aggregate substitution effect of all PC visits reported in Table 3, while an assessment visit produces a much larger substitution effect reducing utilization of ESC by 24.5%. Such a large effect of assessment visits is consistent with their specific aim of reducing the risk of an acute exacerbation of the chronic condition, which may lead to ESC visits. In contrast, a home visit produces a much smaller substitution effect reducing utilization of ESC by 1.1% only. Since home visits are offered to patients who are unable to reach the PC practice, they may also be unable to reach the hospital ED unless their condition deteriorates to the point of requiring an ambulance transfer.

Second, we examine heterogeneity in the substitution effect by individual sociodemographic characteristics by interacting individual income, education, and age with the PC variable. We find no significant socioeconomic gradient in the substitution effect of PC on the demand for ESC (Columns 2 and 3, Table 4), namely, individu-

Table 4 Estimated state dependence and cross effects by some relevant groups

	1		2		3		4	
	PC visit	ESC visit	PC visit	ESC visit	PC visit	ESC visit	PC visit	ESC visit
ESC	−0.0519***	0.1668***	−0.0520***	0.1822***	−0.0521***	0.1820***	−0.0525***	0.1810***
	(0.00)	(0.00)	(0.00)	(0.00)	(0.00)	(0.00)	(0.00)	(0.00)
PC			0.0385***	−0.0881***	0.0421***	−0.0944***	0.0649***	−0.0241***
			(0.00)	(0.00)	(0.00)	(0.00)	(0.00)	(0.01)
Standard visits	0.0452***	−0.1004***						
	(0.00)	(0.00)						
Home visits	0.0384***	−0.0112***						
	(0.00)	(0.00)						
Assessment visit	−0.0255***	−0.2453***						
	(0.00)	(0.01)						
PC × inc. 2nd quintile			0.0040***	−0.0059*				
			(0.00)	(0.00)				
PC × inc. 3rd quintile			0.0056***	−0.0177***				
			(0.00)	(0.00)				
PC × inc. 4th quintile			0.0063***	−0.0039				
			(0.00)	(0.00)				
PC × inc. 5th quintile			0.0127***	−0.0072*				
			(0.00)	(0.00)				
PC × upper sec. edu.					0.0012	−0.0025		
					(0.00)	(0.00)		
PC × tertiary edu.					0.0064***	0.0033		
					(0.00)	(0.00)		
PC × age							−0.0003***	−0.0011***
							(0.00)	(0.00)

Standard errors are in parentheses. *significant at 10%, **significant at 5%, ***significant at 1%

als in different quintiles of the income distribution and different levels of education experience a similar substitution effect. Income and education are associated with a small but statistically significant increment in state dependence in utilization of PC. Finally, we find a significant age gradient in the substitution effect of PC on the demand for ESC (Column 4), which we examine in Sect. 5.3.

5.2 *Latent States Characterization*

The estimated α parameters can be used to characterize the latent states identified by the HMBPo model. Table 5 shows that the propensity of using PC and ESC increases monotonically from States 1 to 3; Fig. 3 adds more details by showing the predicted joint distribution of PC and ESC conditional on the three latent states. This suggests

Table 5 Estimated support points ξ_c for the subject-specific parameters

Latent	PC visit		ESC visit	
State	$\alpha_1(\xi_c)$	SD	$\alpha_2(\xi_c)$	SD
1	−1.7379***	(0.07)	−1.9304***	(0.10)
2	−0.3322***	(0.04)	−0.7052***	(0.05)
3	0.5009***	(0.06)	0.3546***	(0.08)

Standard errors are in parentheses. *significant at 10%, **significant at 5%, ***significant at 1%

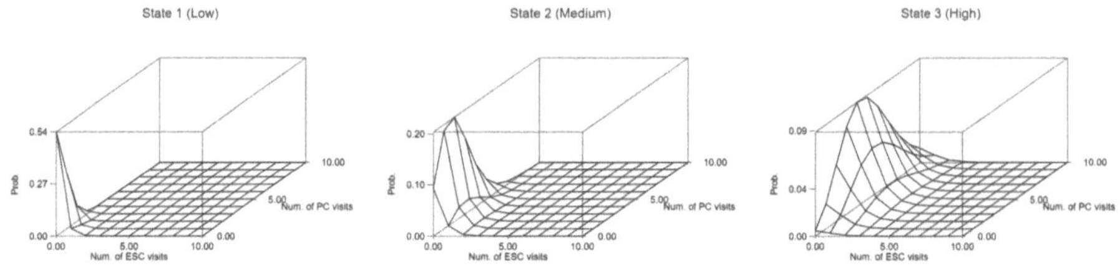

Fig. 3 Predicted joint distribution of PC and ESC visits by latent types

the existence of three states among which individuals can switch over time: the light-user state with a low utilization of both PC and ESC (State 1), the medium-user state with a medium utilization of PC and a low utilization of ESC (State 2), and the heavy-user state with high utilization of both PC and ESC (State 3). Finally, 32% of the population are light users, while medium and heavy users are 52 and 16%, respectively.

We can also characterize the latent states in terms of individual health status. To this end, we can recover the posterior probability that individual i belongs to a latent state k in period t from the HMBPo model. The posterior distribution of the latent process is obtained from

$$q(\xi_i|\tilde{y}_i, x_i, z_i) = \frac{p(\tilde{y}_i|\xi_i, x_i)\,p(\xi_i|z_i)}{p(\tilde{y}_i|x_i, z_i)}, \tag{5}$$

where $\tilde{y}_i$ is the observed pattern of PC and ESC for individual i and ξ_i is a sequence of states. From Eq. (5), we can calculate the posterior probabilities of the different latent states (see Appendix 7), and then their correlation with indicators of individual health status. The left panel of Fig. 4 shows the correlation between the posterior probabilities and three indicators of individual health status, namely, total chronic conditions (in our basket), Charlson index, and total secondary diagnoses. Heavy users are more likely to have a higher Charlson index and a higher number of secondary diagnoses than medium users, who in turn are more likely to have higher scores in these two indicators than light users. Heavy users and medium users have similar probabilities of having additional chronic conditions, but a higher probability than light users. The right panel of Fig. 4 plots the association between number of secondary diagnoses and probability of belonging to a different state. The probability of belonging to the

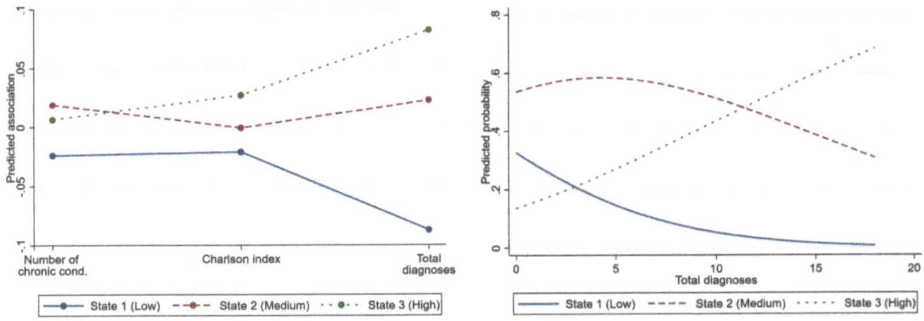

Fig. 4 Association between latent types and health indicators

heavy-user state increases monotonically with the number of secondary diagnoses, while the reverse occurs for the light-user state. Finally, the probability of belonging to the medium-user state is higher than light and heavy users between 1 and 5 secondary diagnoses and then decreases. This supports the interpretation of latent states as indicators of unmeasured individual health, with increasing latent utilization profiles associated with increasing poor health. A similar association is found in other studies leading to a similar interpretation of the latent states (Fooken & Jeet, 2022; Wouterse et al., 2013).

5.3 *Substitution Effect by Age and Latent States*

After describing the latent states, we can move back to the heterogeneity analysis of the substitution effect and expand the age gradient reported in Table 4. Figure 5 plots the average marginal substitution effect in individuals with increasing years of age belonging to different latent states; estimates are obtained from the HMBPo model in Table 4. Heavy users experience a marginal substitution effect that is about three times larger than medium users'. They also experience a marked age gradient with a decreasing substitution effect between ages 40 and 70 and a constant substitution effect thereafter, ranging from -0.1 to -0.08 ESC visits that could be avoided by an additional PC visit. This is a large effect if compared with the average of 0.30 ESC visits per period in the total study population. Also, such a substitution effect shows diminishing marginal decrements in the 40–70 age interval suggesting a higher scope for reducing ESC in the first half of this interval. Moving to the medium users, they show a less marked age gradient than the former with the marginal effect of an additional PC visit ranging from -0.04 to -0.03 ESC visits, which is about 10% of the average ESC utilization in the study population. Finally, individuals in the low-user state show a small marginal substitution effect of -0.01 ESC visits and no age gradient.

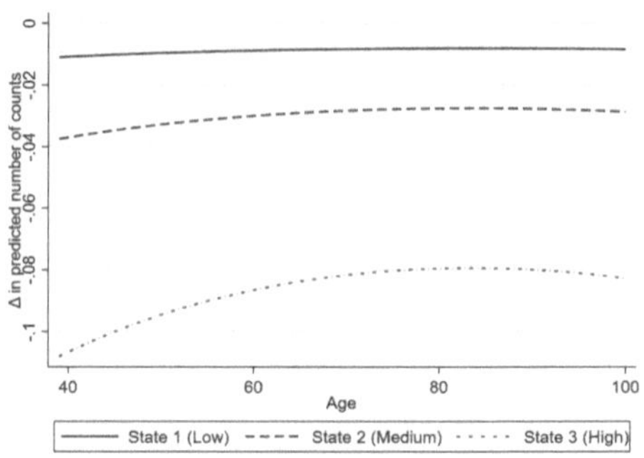

Fig. 5 Estimated average marginal effects of ESC and PC visits

5.4 *Latent States Transitions*

The latent process described in Eq. (2) is also characterized by the transition probabilities in Eq. (4), which can be summarized in the following matrix:

$$\widehat{\Pi} = \begin{bmatrix} 0.89 & 0.10 & 0.01 \\ 0.07 & 0.88 & 0.05 \\ 0.02 & 0.16 & 0.82 \end{bmatrix}.$$

This matrix reports the average probability that an individual switches from one type to another at any time occasion. An individual in the light-user state (first row) is likely to remain in the same state over time with a probability of 89% (first row, first column), while her probability of switching to medium user is 10% (first row, second column), and her probability to become a heavy user is only 1% (first row, last column). A similar picture emerges for the other latent states indicating that latent states are persistent, but not permanent; individuals can switch to their neighbor state with a probability between 5 and 16%, while switching to a more distant state occurs with a probability of 1–2%.[2]

Figure 6 plots the class membership probabilities in every time occasion. Individuals in the medium-user state are more likely to switch to light user in the first 2 years from their initial health shock, while individuals in the heavy-user state share a similar probability of switching state over the whole study period.[3]

The latent state transition probabilities can be modeled as a function of individual sociodemographic characteristics allowing us to examine age dynamics and

[2] We performed a formal test of the hypothesis that individual heterogeneity is time invariant by estimating the HMBPo model under the hypothesis that the transition matrix is diagonal, i.e., absence of time-varying heterogeneity. The likelihood ratio statistic has null asymptotic distribution of $\overline{\chi}^2$ type (Bartolucci, 2006). The log-likelihood for the restricted model is equal to -1127474.25 which, compared to that one of the unrestricted models reported in Table 2, leads to a likelihood ratio statistic with 6 dof rejecting the null hypothesis of time-constant heterogeneity with a p-value $< 10^{-3}$.

[3] The whole study period consists of 20 years quarters from 2012 to 2017; the first two quarters are used to estimate the HMBPo initial conditions and to allow for the time lag in state dependence variables; hence, the number of quarters plotted in Fig. 6 is 18.

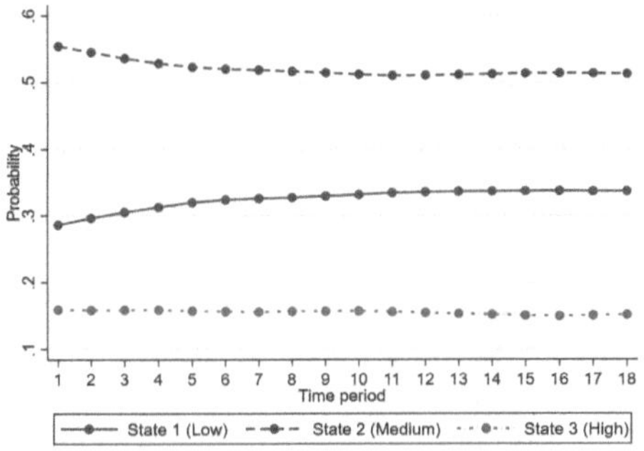

Fig. 6 Estimated average probabilities of latent states in each period

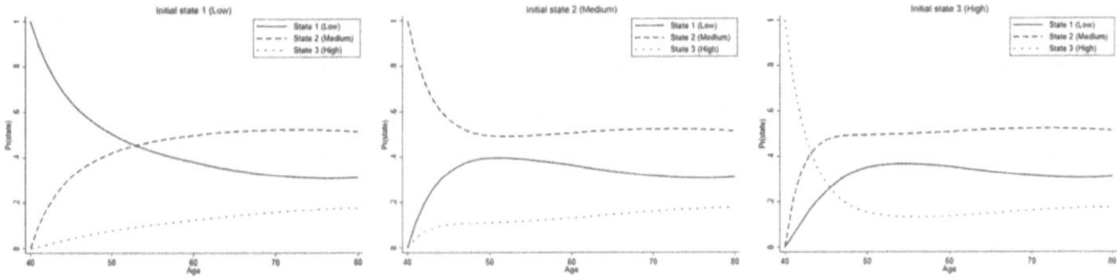

Fig. 7 Transition probabilities of being in a state calculated up to age 80 for individuals in a particular initial state at age 40

socioeconomic inequalities in switching latent states over time. First, we examine the probabilities of changing state over the lifespan of a 40-year-old individual who starts in one of the three alternative states: light-user, medium-user, and heavy-user states. Using the backward recursion procedure described in Appendix 7, we can compute the probability that this individual will be in state c at age 41 given her initial state in the previous year and her transition probabilities (Eq. (4)). This procedure is repeated up to age 80 and results are plotted in Fig. 7.

In each of the initial states examined, the probability of changing state increases rapidly between 40 and 50 years old and stabilizes thereafter. However, individuals starting as medium and heavy users (second and third panels of Fig. 7) experience a more rapid change of state in the 40–50 age interval than individuals starting as light users (first panel of Fig. 7). This can be interpreted as a positive change as about 50% of the initially medium and heavy users switch to light users in this age interval, while about 50% of the initially light users switch to medium users. In the 50–80 age interval, light users continue the trajectories started in the previous interval, although at a slower pace than in the former interval (first panel of Fig. 7). In contrast, medium and heavy users experience a change of trajectory with a progressive reduction in the probability of moving to the light-user state (second and third panels of Fig. 7). This suggests that health improvements and changes in preferences for utilization of care are likely to occur in the 40–50 age interval and individuals in this interval could be more responsive to policy interventions than older individuals.

Overall, individuals starting as light users at 40 years old have about a 30% probability of remaining in their initial state when they reach 80 years old, and a 50% probability of becoming medium users, and a 20% probability of becoming heavy users. In contrast, individuals starting as medium users have about a 50% probability of remaining in their initial state, a 30% probability of becoming light users, and a 20% probability of becoming heavy users when they reach 80 years old. Finally, heavy users have a 20% probability of remaining in their initial state, a 30% probability of becoming light users, and a 50% probability of becoming medium users when they reach 80 years old. Overall, the probability of switching from a lower to a higher user state and the probability of remaining in the heavy-user state increase monotonically with age in each of the initial states examined from age 50. These age trajectories suggest a progressive deterioration of health in individuals with a chronic condition and an increment in utilization of care as they become older (De Meijer et al., 2011; Howdon & Rice, 2018).

Second, we examine heterogeneity in the age trajectories of the latent states by comparing the trajectories of individuals who are socioeconomically deprived with the rest of the population. We define the former as individuals from the bottom quintile of the income distribution with a compulsory level of education and living alone, hence with less access to informal care from a family member. Figure 8 shows that the latent states of deprived individuals follow very similar trajectories than the rest of the population. The most noticeable difference seems to emerge in deprived individuals starting in the heavy-user state (panel 5, Fig. 8) who experience a more rapid transition from their initial state to the medium-user and light-user states in the 40–50 age interval as compared to non-deprived individuals starting in the same state (panel 6, Fig. 8). Moreover, the former experience a noticeable change in the trajectory of their latent states in the 50–80 age interval with a progressive reduction in the probability of moving to the light-user state and a progressive increment in the probability of moving back to the heavy-user state, while the latter experience flat trajectories in the same age interval. This may suggest that deprived individuals starting in a poor health state may improve faster when they are relatively young, but are more likely to regress back to a poor health state when they are older.

6 Discussion and Conclusion

We study the substitution effect between PC and ESC in patients with chronic conditions using a structural model of the demand of PC and ESC. The HMBPo allows us to model three distinct sources of correlation characterizing the data-generating process: correlation within and between outcomes, correlation from unmeasured heterogeneity in patient health and preferences for utilization, and correlation from time-specific shocks that may affect the demand of both PC and ESC. The model assumes that the data-generating process follows a first-order Markov Chain, namely, that utilization in the present period depends on utilization in the past period after allowing for observable covariates (Wouterse et al., 2013). Our analysis extends the

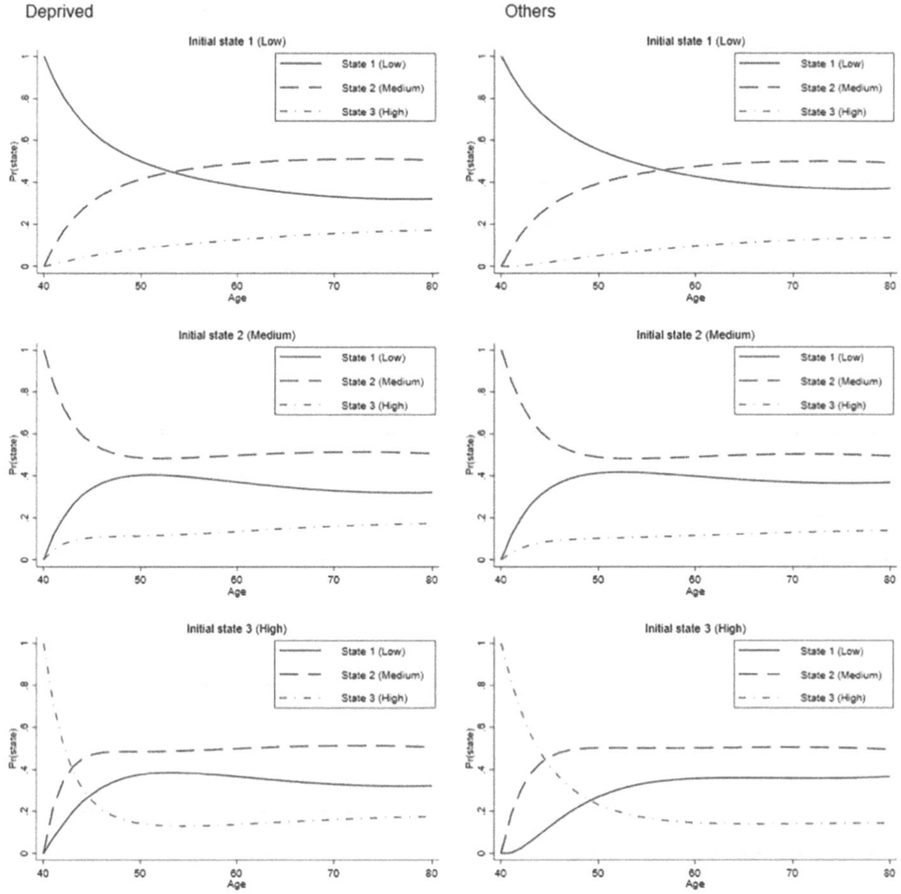

Fig. 8 Transition probabilities of being in a state calculated up to age 80 for individuals in a particular initial state at age 40 and different level of deprivation

scope of existing studies that identify the substitution effect from the exogenous variation in PC generated by the introduction of a new policy. In particular, our analysis of the substitution effect is not constrained to a policy period, policy targeted areas, and physicians' response to policy incentives. Indeed, our approach is oriented toward producing evidence that may inform new policies, rather than evaluating the effect of existing policies, and contribute to the existing literature in this direction.

We find a noticeable substitution effect with an additional PC visit reducing utilization of ESC by 9.5%; our findings are consistent with evidence from other studies (Dolton & Pathania, 2016; Fortney et al., 2005; Iezzi et al., 2014; Pinchbeck, 2019; Whittaker et al., 2016). Notice that, under our maintained structural assumptions—namely, the joint dynamic process between PC and ESC, sequential exogeneity given past outcomes and latent states, no remaining within-period shocks affecting PC and ESC beyond the common shock we model, and accurate measurement—the estimated -9.5% can be interpreted as a causal semi-elasticity within the model. This is a structural (model-based) notion of causality; it does not rely on quasi-experimental variation and therefore should not be read as a policy-induced causal effect. Policy-counterfactual claims would require exogenous shifts in primary care supply or access.

We also find that some PC visits are more effective than others in reducing utilization of ESC, namely, the substitution effect of a preventive assessment visit is three times the effect of a standard consultation, while home visits produce a much smaller substitution effect. Preventive assessment visits aim at reducing the risk of acute exacerbation of the chronic condition by assessing the progress of the disease and comorbidities and suggesting behavioral changes and pharmaceutical therapies when appropriate. New policies aiming at reducing utilization of unnecessary ESC could have a greater impact by supporting the supply of preventive assessment visits in PC. In 2018, Denmark has introduced a differentiated capitation payment covering all services delivered to patients with diabetes and COPD in an effort to control the annual cost of these patients. However, this policy might reduce the monetary incentive for delivering PC services to these patients, thus resulting in a higher utilization of ESC if patients substitute the former with the latter.

From a policy perspective, the stronger substitution effect of preventive assessment visits (vs. standard or home visits) points to three practical levers: (i) align incentives to increase and target assessments for chronic patients (e.g., small risk-adjusted top-ups or light P4P tied to annual assessments and post-ED reviews); (ii) improve access and continuity on the substitution margin (same-/next-day slots for exacerbations, proactive recall after ED, after-hours GP cooperatives); and (iii) embed team-based chronic care (nurse checks, pharmacist medication reviews, brief training on action plans).

We also perform a heterogeneity analysis of the substitution effect by individual sociodemographic characteristics, including age, income, and education. We find that the substitution effect decreases with age with diminishing marginal increments in the 40–70 age interval, and it is much larger in individuals with a heavy-user profile than in individuals with medium- and light-user profiles. In contrast, we find no evidence of a socioeconomic gradient in the substitution effect. This suggests that future health policies may have a stronger impact by targeting PC services delivered to middle-aged individuals with a chronic condition and who are already frequent users of health services. Also, the scope for reducing unnecessary ESC by targeting PC services delivered to patients with a low socioeconomic status might be no different than in other patients in a universal healthcare system in which access to care is free of charge at the point of demand. Within Denmark's universal system—free at the point of use with strong GP gatekeeping—financial barriers and access pathways are largely equalized, plausibly compressing SES differences specifically on the substitution margin between primary care and emergency secondary care. In other words, while SES correlates with overall utilization and health profiles, once patients are "in the system" the preventive management that averts ED-treated exacerbations appears to operate similarly across SES groups. This interpretation aligns with our estimates showing SES main effects on utilization but non-significant PC×SES interactions in the ESC equation, and with the institutional setting described in Appendix 8. That said, this pattern may be system-contingent: in settings with cost-sharing, weaker gatekeeping, or larger within-area supply disparities, SES gradients in substitution could emerge through affordability, navigation costs, or differential care coordination. We therefore view the absence of SES heterogeneity as consistent with Denmark's institutional design, and we caution that transferability to non-universal systems should be empirically verified.

Some remarks on external validity and generalizability of our estimates are also needed. Particularly, estimates are obtained in Denmark, a tax-funded universal system with free access at the point of use and strong GP gatekeeping; these features dampen financial barriers and shape pathways between primary and emergency secondary care. The behavioral mechanisms we study—preventive management in primary care averting acute exacerbations that trigger ED presentations—are not system-specific and should carry over to similar universal systems (e.g., other Nordic countries, the UK, Italy, Spain), although magnitudes will vary with payment schedules, access rules, and ED organization. We expect qualitatively similar substitution patterns for other highly prevalent chronic conditions where ambulatory management can avert decompensations (e.g., heart failure) or treatment-related complications (e.g., oncology), but condition-specific trajectories (severity, relapse risk, monitoring intensity) may modulate the size of effects. We therefore view our results as externally relevant for universal systems with comparable GP roles, while recommending system- and condition-specific validation when financial incentives or access pathways differ materially.

Finally, we performed a heterogeneity analysis of the latent states and their trajectories by age and socioeconomic status. Latent states describe individuals' utilization profiles and are associated with individuals' unmeasured health and preferences for utilization (Harris et al., 2018). Although latent states are persistent over time, they are not permanent, and individuals can switch to a neighbor state over time. Moreover, we find that probability of changing state decreases with age. Switching to a different utilization profile is likely to occur in the 40–50 age interval, in contrast latent states are relatively stable in older ages with their trajectories pointing to a higher utilization profile in the 70–80 age interval. We did not find noticeable differences in the trajectories of the latent states by socioeconomic status. This suggests that aging might be associated with a deterioration of unmeasured health and a movement toward higher utilization profiles in individuals with chronic conditions (Lafortune et al., 2009; Parker & Thorslund, 2007).

7 Appendix: Marginal and Transition Probabilities

In order to compute efficiently the densities and probabilities, the Baum–Welch recursions are usually employed. Let $\boldsymbol{w}_{it}$ collect the full set of covariates used to model the latent process ($\boldsymbol{z}_{it}$), PC and ESC ($\boldsymbol{x}_{it}$, $\boldsymbol{y}_{it-1}$) The first recursion introduced by Baum and Welch is the forward recursion, which is usually employed to efficiently compute the individual log-likelihood ℓ_i. To implement the forward recursion let

$$f^{(t)}(\xi_c, \widetilde{\boldsymbol{y}}|\boldsymbol{w}) = p(\boldsymbol{\alpha}_{it} = \boldsymbol{\xi}_c, \boldsymbol{y}_1, \ldots, \boldsymbol{y}_t|\boldsymbol{w}_{i1}, \ldots, \boldsymbol{w}_{it}),$$

with $t = 1, \ldots, T$. The recursion is usually initialized in the first period as follows:

$$f^{(1)}(\xi_c, \widetilde{\boldsymbol{y}}|\boldsymbol{w}) = p^{(1)}\big(\boldsymbol{\alpha}_{i1} = \boldsymbol{\xi}_c|z_{i1}, \boldsymbol{y}_{i0}\big)p(\boldsymbol{y}_{i1}|\boldsymbol{\alpha}_{i1}, \boldsymbol{x}_{i1}, \boldsymbol{y}_{i,0})$$

and proceed for the subsequent $t = 2, \ldots, T_i$ periods as follows:

$$f^{(t)}(\xi_c, \widetilde{\boldsymbol{y}}|\boldsymbol{w}) = \sum_{\overline{c}=1}^{k} f^{(t)}(\xi_{\overline{c}}, \xi_c, \widetilde{\boldsymbol{y}}|\boldsymbol{w}),$$

with $c = 1, \ldots, k$ and where

$$\begin{aligned} f^{(t)}(\xi_{\overline{c}}, \xi_c, \widetilde{\boldsymbol{y}} \mid \boldsymbol{w}) = f^{(t-1)}(\xi_{\overline{c}}, \widetilde{\boldsymbol{y}} \mid \boldsymbol{w})\, p^{(t)}\big(\boldsymbol{\alpha}_{it} = \boldsymbol{\xi}_{\overline{c}} \mid \boldsymbol{\alpha}_{i,t-1} = \boldsymbol{\xi}_c, z_{it}\big) \\ \times\, p(\boldsymbol{y}_{it} \mid \boldsymbol{\alpha}_{it}, \boldsymbol{x}_{it}, \boldsymbol{y}_{i,t-1}). \end{aligned}$$

Therefore at the end of the recursion the individual log-likelihood is given by

$$\ell_i = \log\left\{\sum_{c=1}^{k} f^{(T_i)}(\xi_c, \widetilde{\boldsymbol{y}}|\boldsymbol{w})\right\}.$$

To compute the densities and probabilities used to obtain the posterior distribution of every latent state and of every pair of consecutive latent states, we employ the Baum and Welch's backward recursion. This recursion starts with $b^{(T)}(\widetilde{\boldsymbol{y}}|\boldsymbol{w}) = 1$, while in the subsequent periods:

$$b^{(t)}(\widetilde{\boldsymbol{y}}|\boldsymbol{w}) = p(\boldsymbol{y}_{t+1}, \ldots, \boldsymbol{y}_{T_i}|\boldsymbol{\alpha}_{it} = \boldsymbol{\xi}_{\overline{c}}, \boldsymbol{w}_{it+1}, \ldots, \boldsymbol{w}_{iT_i}),$$

with $t = 1, \ldots, T_i$ and $\overline{c} = 1, \ldots, k$. At each step of the backward recursion and starting from $t = T_i - 1$ to $t = 1$ the following is computed:

$$b^{(t)}(\widetilde{\boldsymbol{y}}|\xi_c, \boldsymbol{w}) = \sum_{c=1}^{k} b^{(t)}(\xi_c, \widetilde{\boldsymbol{y}}|\xi_{\overline{c}}, \boldsymbol{w}),$$

where

$$\begin{aligned} b^{(t)}(\xi_c, \widetilde{\boldsymbol{y}} \mid \xi_{\overline{c}}, \boldsymbol{w}) = b^{(t+1)}(\widetilde{\boldsymbol{y}} \mid \xi_c, \boldsymbol{w})\, p^{(t+1)}\big(\boldsymbol{\alpha}_{it+1} = \boldsymbol{\xi}_c \mid \boldsymbol{\alpha}_{i,t} = \boldsymbol{\xi}_{\overline{c}}, z_{it}\big) \\ \times\, p(\boldsymbol{y}_{it+1} \mid \boldsymbol{\alpha}_{it+1}, \boldsymbol{x}_{it+1}, \boldsymbol{y}_{i,t}). \end{aligned}$$

In particular, the probabilities to observe that observation i belongs to a latent state k at period t are given by

$$q^{(t)}\left(\xi_c|\widetilde{y}_i, w_i\right)= \frac{f^{(t)}(\xi_c, \widetilde{y}|w)b^{(t)}(\widetilde{y}|w)}{\sum_{c=1}^{k} f^{(t)}(\xi_c, \widetilde{y}|w)},$$

with $t = 1, \dots, T_i$, while the probabilities to observe that observation i switches from latent state c to $\overline{c}$ at time t are given by

$$q^{(t)}\left(\xi_{\overline{c}}, \xi_c|\widetilde{y}_i, w_i\right)= \frac{f^{(t)}(\xi_{\overline{c}}, \xi_c, \widetilde{y}|w)b^{(t)}(\widetilde{y}|w)}{\sum_{c=1}^{k} f^{(t)}(\xi_c, \widetilde{y}|w)},$$

with $t = 2, \dots, T_i$.

8 Appendix: Institutional Framework

8.1 Primary Care

The Danish National Health System (NHS) offers a universal coverage to residents in Denmark and access to PC and ESC services is free of charge at the point of demand. PC physicians are independent professionals contracted by the Danish NHS under a mix-remuneration system with 70% of their income generated by capitation fees, namely, a fix payment for each patient enrolled in their list, while the remaining 30% is generated by fee-for-service payments. The latter include a fee for standard face-to-face consultations (about 19 Euros in 2017) during which the PC physician can examine a single health issue reported by the patient, and preventive assessment visits (50 Euros) that PC physicians are recommended to offer to their patients with chronic conditions at least once a year. During the assessment visit, the PC physician evaluates the progress of the chronic condition and comorbidities and agrees a set of preventive efforts with the patient, including behavioral changes and pharmaceutical therapies aiming at reducing the risk of acute exacerbation of the disease. Finally, home visits are offered at the discretion of the PC physician to those patients who require medical assistance but are too sick to reach the PC practice; home visits are remunerated by a separate fee which is proportional to the distance covered (about 39 Euros on average).

8.2 Emergency Secondary Care

ESC services are provided by 21 large multi-service hospitals, which are non-profit public organizations serving a local population of 250,000 residents and managed by the five Danish Regions. While patients need a PC referral to access elective secondary care, they can access ESC directly by calling an emergency number managed by the Emergency Medical Coordination Centers that direct them to the closest hospital with an ED available. Hospital services are reimbursed by the Danish Regions through a system of DRG tariffs centrally determined by the Department of Health on the basis of the average costs reported by hospitals every year. DRGs were initially introduced in 2003 as a tool to incentivize hospital productivity and in 2005 were officially adopted as a reimbursement system for hospital services. During the period of our study, the remuneration of both PC and ESC services remained stable as well as the total number of hospitals with an emergency department.

Ethics statement The analysis involved the use of previously collected and pseudonymized individual-level data on access to health services. Ethics approval to conduct the study has been granted by Statistics Denmark (DST) and the University Research and Ethics Commission.

References

Alfò, M., & Maruotti, A. (2010). Two-part regression models for longitudinal zero-inflated count data. *Canadian Journal of Statistics, 38*(2), 197–216.

Atella, V., & Deb, P. (2008). Are primary care physicians, public and private sector specialists substitutes or complements? Evidence from a simultaneous equations model for count data. *Journal of Health Economics, 27*, 770–785.

Bago d'Uva, T. (2005). Latent class models for use of primary care: Evidence from a British panel. *Health Economics, 14*(9), 873–892.

Bago d'Uva, T., & Jones, A. M. (2009). Health care utilisation in Europe: New evidence from the ECHP. *Journal of Health Economics, 28*, 265–279.

Baier, N., Geissler, A., Bech, M., Bernstein, D., Cowling, T. E., Jackson, T., van Manen, J., Rudkjøbing, A., & Quentin, W. (2019). Emergency and urgent care systems in Australia, Denmark, England, France, Germany and the Netherlands - Analyzing organization, payment and reforms. *Health Policy, 123*, 1–10.

Bartolucci, F. (2006). Likelihood inference for a class of latent Markov models under linear hypotheses on the transition probabilities. *Journal of the Royal Statistical Society: Series B (Statistical Methodology), 68*, 155–178.

Bartolucci, F., & Farcomeni, A. (2015). Information matrix for hidden Markov models with covariates. *Statistics and Computing, 25*, 515–526.

Contoyannis, P., Jones, A., & Rice, N. (2004). The dynamics of health in the British household panel survey. *Journal of Applied Econometrics, 19*, 473–503.

De Meijer, C., Koopmanschap, M., d'Uva, T. B., & Van Doorslaer, E. (2011). Determinants of long-term care spending: Age, time to death or disability? *Journal of Health Economics, 30*, 425–438.

Dempster, A. P., Laird, N. M., & Rubin, D. B. (1977). Maximum likelihood from incomplete data via the EM algorithm. *Journal of the Royal Statistical Society. Series B (Methodological), 39*, 1–38.

Dolton, P., & Pathania, V. (2016). Can increased primary care access reduce demand for emergency care? Evidence from England's 7-day GP opening. *Journal of Health Economics, 49*, 193–208.

Doyle, J. J. (2011). Returns to local-area health care spending: Evidence from health shocks to patients far from home. *American Economic Journal: Applied Economics, 3*, 221–43.

Dusheiko, M., Gravelle, H., Martin, S., Rice, N., & Smith, P. C. (2011). Does better disease management in primary care reduce hospital costs? Evidence from English primary care. *Journal of Health Economics,30*, 919–932.

Fooken, J., & Jeet, V. (2022). Using Australian panel data to account for unobserved factors in measuring inequities for different channels of healthcare utilization. *The European Journal of Health Economics, 23*, 717–728.

Fortney, J. C., Steffick, D. E., Burgess, J. F., Jr., Maciejewski, M. L., & Petersen, L. A. (2005). Are primary care services a substitute or complement for specialty and inpatient services? *Health Services Research, 40*, 1422–1442.

García-Gómez, P., Van Kippersluis, H., O'Donnell, O., & Van Doorslaer, E. (2013). Long-term and spillover effects of health shocks on employment and income. *Journal of Human Resources, 48*, 873–909.

Gravelle, H., Sutton, M., & Ma, A. (2010). Doctor behaviour under a pay for performance contract: Treating, cheating and case finding? *The Economic Journal, 120*, F129–F156.

Harris, P., Whitty, J. A., Kendall, E., Ratcliffe, J., Wilson, A., Littlejohns, P., & Scuffham, P. A. (2018). The importance of population differences: Influence of individual characteristics on the Australian public's preferences for emergency care. *Health Policy, 122*, 115–125.

Heckman, J. J. (1981). Heterogeneity and state dependence. In S. Rosen (Ed.), *Studies in labor markets, Chapter 3* (pp. 91–140). Chicago, IL: University of Chicago Press.

Howdon, D., & Rice, N. (2018). Health care expenditures, age, proximity to death and morbidity: Implications for an ageing population. *Journal of Health Economics, 57*, 60–74.

Hyppolite, J., & Trivedi, P. (2012). Alternative approaches for econometric analysis of panel count data using dynamic latent class models (with application to doctor visits data). *Health Economics, 21*(Suppl. 1), 101–128.

Iezzi, E., Lippi Bruni, M., & Ugolini, C. (2014). The role of GP's compensation schemes in diabetes care: Evidence from panel data. *Journal of Health Economics, 34*, 104–120.

Kohn, J. L., & Liu, J. S. (2013). The dynamics of medical care use in the British household panel survey. *Health Economics, 22*, 687–710.

Kollerup, A., Kjellberg, J., & Ibsen, R. (2022). Ageing and health care expenditures: the importance of age per se, steepening of the individual-level expenditure curve, and the role of morbidity. *The European Journal of Health Economics, 23*, 1121–1149.

Lafortune, L., Béland, F., Bergman, H., & Ankri, J. (2009). Health status transitions in community-living elderly with complex care needs: A latent class approach. *BMC Geriatrics, 9*, 1–14.

Laudicella, M., Donni, P. L., & Prete, V. (2024). Healthcare utilisation by diabetic patients in Denmark: The role of primary care in reducing emergency visits. *Health Policy, 145*, Article 105079.

Laudicella, M., & Li Donni, P. (2022). The dynamic interdependence in the demand of primary and emergency secondary care: A hidden Markov approach. *Journal of Applied Econometrics, 37*, 521–536.

Laudicella, M., Li Donni, P., Olsen, K. R., & Gyrd-Hansen, D. (2022). Age, morbidity, or something else? A residual approach using microdata to measure the impact of technological progress on health care expenditure. *Health Economics, 31*, 1184–1201.

Lippi Bruni, M., Mammi, I., & Ugolini, C. (2016). Does the extension of primary care practice opening hours reduce the use of emergency services? *Journal of Health Economics, 50*, 144–155.

Oakes, D. (1999). Direct calculation of the information matrix via the EM. *Journal of the Royal Statistical Society: Series B (Statistical Methodology), 61*, 479–482.

OECD. (2015). *Fiscal Sustainability of Health Systems*. Paris: OECD Publishing.

Parker, M. G., & Thorslund, M. (2007). Health trends in the elderly population: Getting better and getting worse. *The Gerontologist, 47*, 150–158.
Pinchbeck, E. W. (2019). Convenient primary care and emergency hospital utilisation. *Journal of Health Economics, 68*, Article 102242.
Pulleyblank, R., Skovsgaard, C., & Olsen, K. R. (2020). New primary care incentive scheme for type 2 diabetes in Denmark. *Value in Health, 23*, S408–S409.
Reynolds, R., Dennis, S., Hasan, I., Slewa, J., Chen, W., Tian, D., Bobba, S., & Zwar, N. (2018). A systematic review of chronic disease management interventions in primary care. *BMC Family Practice, 19*, 1–13.
Riphahn, R. T., Wambach, A., & Million, A. (2003). Incentive effects in the demand for health care: A bivariate panel count data estimation. *Journal of Applied Econometrics, 18*, 387–405.
Rudkjøbing, A., Vrangbaek, K., Birk, H. O., Andersen, J. S., & Krasnik, A. (2015). Evaluation of a policy to strengthen case management and quality of diabetes care in general practice in Denmark. *Health Policy, 119*, 1023–1030.
Thygesen, L. C., C. Daasnes, I. Thaulow, & H. Brønnum-Hansen (2011). Introduction to Danish (nationwide) registers on health and social issues: Structure, access, legislation, and archiving. *Scandinavian Journal of Public Health, 39*, 12–16.
Tsiachristas, A., Dikkers, C., Boland, M. R., & Rutten-van Mölken, M. P. (2013). Exploring payment schemes used to promote integrated chronic care in Europe. *Health Policy, 113*, 296–304.
Tummalapalli, S. L., Estrella, M. M., Jannat-Khah, D. P., Keyhani, S., & Ibrahim, S. (2022). Capitated versus fee-for-service reimbursement and quality of care for chronic disease: A US cross-sectional analysis. *BMC Health Services Research, 22*, 1–12.
Walsh, B., & Laudicella, M. (2017). Disparities in cancer care and costs at the end of life: Evidence from England's national health service. *Health Affairs, 36*, 1218–1226.
Welch, L. R. (2003). Hidden Markov models and the Baum-Welch algorithm. *IEEE Information Theory Society Newsletter, 53*, 10–13.
Whittaker, W., Anselmi, L., Kristensen, S. R., Lau, Y. -S., Bailey, S., Bower, P., Checkland, K., Elvey, R., Rothwell, K., & Stokes, J. (2016). Associations between extending access to primary care and emergency department visits: A difference-in-differences analysis. *PLoS Medicine, 13*, Article e1002113.
Wooldridge, J. M. (2005). Simple solutions to the initial conditions problem in dynamic, nonlinear panel data models with unobserved heterogeneity. *Journal of Applied Econometrics, 20*, 39–54.
Wouterse, B., Huisman, M., Meijboom, B. R., Deeg, D. J., & Polder, J. J. (2013). Modeling the relationship between health and health care expenditures using a latent Markov model. *Journal of Health Economics, 32*, 423–439.

The Demand of Primary and Secondary Care: A Bayesian Hierarchical Approach

Paolo Li Donni and Catia Nicodemo

Abstract Health systems increasingly face the dual challenge of rising demand and persistent inequalities in access to care. Understanding how patients use different services—particularly the balance between primary and secondary care—is essential for informing policy. This study develops a Bayesian hierarchical bivariate Poisson model to jointly analyze general practitioner consultations and emergency department visits among older adults newly diagnosed with hypertension, using data from the UK Clinical Practice Research Datalink. The framework accounts for both individual-level heterogeneity and practice-level fixed effects, capturing substitution or complementarity between services. Results show that multimorbidity strongly drives both utilization, while socioeconomic deprivation is mainly associated with higher emergency care use, pointing to persistent inequalities. We also find substantial heterogeneity across GP practices, suggesting that organizational and contextual factors significantly shape care pathways.

Keywords Primary care · Secondary care · Chronic conditions · Hidden Markov models · Panel data

1 Introduction

Health systems increasingly face the dual challenge of rising demand and persistent inequalities in access to care. Understanding how patients use different services—particularly the balance between primary and secondary care—is central for design-

---

Supplementary Information The online version contains supplementary material available at https://doi.org/10.1007/978-3-032-24141-2_8.

P. Li Donni (✉)
University of Palermo, Palermo, Italy
e-mail: paolo.lidonni@unipa.it

C. Nicodemo
Brunel University of London, London, England
e-mail: catia.nicodemo@brunel.ac.uk

University of Oxford, Oxford, England

F. Bartolucci et al. (eds.), *Models for Longitudinal Data with Applications to Early Warning Systems*, Springer Proceedings in Mathematics & Statistics 553,
https://doi.org/10.1007/978-3-032-24141-2_8

ing equitable and efficient policies. Traditional single-outcome analyses often fail to capture these interdependencies, as use of one service may substitute or complement use of another (Cameron & Johansson, 1998; Chib & Winkelmann, 2001; Munkin & Trivedi, 1999). Bivariate and multivariate count models provide a robust alternative by allowing correlated outcomes to be studied jointly (Karlis, 2006; Khafrim et al., 2008). Recent applications further highlight their value in policy evaluation, capturing substitution, complementarity, and disparities in healthcare utilization (Kouakou et al., 2021; Qarmalah & Alzaid, 2023).

A central methodological challenge is that healthcare utilization data are typically recorded as counts and these outcomes are often interdependent: for instance, increased access to general practitioner (GP) consultations may reduce or increase reliance on emergency department (ED) visits. To address this complexity, a wide range of bivariate and multivariate Poisson models have been developed (Berkhout & Plug, 2004; Karlis, 2006; Khafrim et al., 2008). Recent extensions, such as doubly inflated or zero-inflated formulations (Kouakou et al., 2021; Qarmalah & Alzaid, 2023; Sengupta et al., 2016), further enhance their flexibility to handle excess zeros, matched counts, and negative correlations that are common in administrative health data.

Beyond technical refinements, researchers have increasingly turned to hierarchical models that integrate both individual- and provider-level effects. These approaches make it possible to disentangle whether patterns of service use primarily reflect patient-level needs and behaviors (e.g., multimorbidity, deprivation, health-seeking behavior) or organizational and contextual differences across practices.

This distinction is not only methodological but also highly policy-relevant: if most variation is explained by practice structure, interventions might focus on redistributing resources, revising staffing, or reorganizing care delivery. By contrast, if patient-level drivers dominate, policy attention must shift toward prevention, chronic disease management, and strategies that change how individuals engage with the health system. Bayesian frameworks provide an especially powerful way to model such complex structures, yielding robust inference while incorporating uncertainty (Karlis, 2006; Khafrim et al., 2008; Mohammadi et al., 2016; Sengupta et al., 2016). Their versatility is illustrated by applications across a wide range of domains—from drug use (Cameron & Johansson, 1998), health behaviors (Mohammadi et al., 2016), and fertility (Mponda & Kaombe, 2024), to service utilization studies in Australia (Gonçalves et al., 2025).

In this study, we extend this literature by developing a Bayesian hierarchical bivariate Poisson model to jointly analyze GP consultations and ED visits among older adults newly diagnosed with hypertension, using linked data from the UK Clinical Practice Research Datalink (CPRD) and the Index of Multiple Deprivation (IMD).

There are two main contributions of this paper to literature. First, we extend bivariate Poisson model, to the hierarchical case, where multiple effects at GP practice levels are estimated in suitable Bayesian framework. Second, by including these effects this framework allows us to disaggregate the drivers of utilization into those attributable to patients and those shaped by GP practice structures. The framework incorporates individual-level heterogeneity and practice-level fixed effects, capturing both patient and organizational influences on service use.

Our results indicate that multimorbidity is a key determinant of healthcare utilization, driving higher use of both GP consultations and ED visits. Socioeconomic deprivation, by contrast, shows a stronger association with increased ED use, underscoring persistent inequalities in access and outcomes. We also identify notable heterogeneity across GP practices, suggesting that organizational and contextual factors play a role in shaping care pathways. However, the estimated correlation between GP and ED use at the practice level is weak but positive, pointing to only limited substitution between services.

From a policy perspective, this distinction is critical. Much current debate emphasizes expanding GP capacity as a way to reduce hospital pressure. However, our results suggest that such supply-side measures, while potentially beneficial, should be accompanied with policy addressing patient behaviors and needs.

This chapter is organized as follows: Sect. 2 reviews the database used in the empirical estimation. Section 3 introduces the Bayesian hierarchical extension. Section 4 describes the estimation model. Section 5 discusses empirical findings, while Sect. 6 discusses model's strengths and limitations with some concluding remarks. To implement the model described in Sect. 3, we rely on a set of user-written routines in `R` specifically designed to structure and preprocess the data for hierarchical bivariate count models. These routines, which are available online, are integrated with a custom `Stan` program, called via the `cmdstanr` interface, to perform posterior sampling, parallelization, and inference.

2 Data Source and Study Population

For this study, we utilized data from the Aurum Clinical Practice Research Datalink (CPRD),[1] focusing on patients aged 50 years and older over the period from 2008 to 2017. The Aurum CPRD is one of the largest and most comprehensive databases of anonymized primary care electronic health records in the United Kingdom. It includes data from approximately 800 general practices, covering around 7% of the UK population, which amounted to over 3 million registered patients by 2017. Importantly, previous validation studies have demonstrated that the Aurum CPRD cohort is broadly representative of the wider UK population in terms of key demographic characteristics (Herrett et al., 2015; Mathur et al., 2014), supporting the generalizability of findings derived from this resource.

To enrich our analysis and gain a more holistic understanding of patient health trajectories, we leveraged CPRD's robust data linkage capabilities. This enabled integration of multiple complementary data sources spanning the same study period. We incorporated the Index of Multiple Deprivation (IMD), a widely used area-level socioeconomic measure in the UK. The IMD aggregates indicators across various

[1] For more information on CPRD data, see https://www.cprd.com/.

domains, including income, employment, education, health, crime, housing, and living environment, to generate deprivation rankings for small geographic areas. Incorporating the IMD allowed us to contextualize individual healthcare utilization patterns within broader socioeconomic environments, thereby facilitating analyses of inequalities and disparities in primary care access and use.

Our research specifically targeted patients diagnosed with hypertension—a highly prevalent chronic condition that requires ongoing management and monitoring within primary care settings. Hypertension affects approximately 30% of adults globally (World Health Organization, 2021) and imposes substantial clinical and economic burdens on healthcare systems worldwide. Due to its progressive nature and established role as a leading risk factor for cardiovascular disease and related complications (Whelton et al., 2018), timely and effective primary care management of hypertension is critical to improving patient outcomes.

We constructed a cohort of incident hypertension cases by identifying individuals who received a new hypertension diagnosis in 2011, ensuring no prior recorded diagnosis of hypertension in the preceding 2 years (2009–2010). This approach allowed us to focus on newly diagnosed patients and track the natural progression of their condition over time. We followed these patients longitudinally through their electronic health records until 2017, capturing information on subsequent diagnoses, treatments, and healthcare utilization patterns.

The primary outcomes of interest were measured at the patient-year level and comprised of two count variables: (1) the total number of primary care visits, reflecting overall engagement with general practice services, and (2) the number of consultations in the emergency department (ED). This distinction enabled us to explore the intensity of care delivery and the potential for substitution or complementarity between consultations with general practitioners and ED.

We use a rich set of covariates to adjust for confounding factors influencing healthcare use. At the individual level, we included demographic variables, such as age and gender, as well as a health status indicator based on the count of coexisting chronic conditions (comorbidities). Recognizing the persistence of healthcare utilization behaviors over time, lagged values of both general practitioner and ED consultations from the previous year ($t - 1$) were incorporated to capture temporal dependencies. Socioeconomic context was controlled through the inclusion of the IMD.

The analytical sample comprised of all patients aged between 50 and 97 years who were actively registered with a participating practice during the study period. To reduce the influence of extreme values and potential data entry errors, observations with healthcare utilization above the 99th percentile were excluded from the analysis. This trimming helped ensure robust and reliable inference while preserving the representativeness of typical utilization patterns.

Together, these data sources and study design elements provided a comprehensive and nuanced framework to investigate primary care utilization patterns among older adults with newly diagnosed hypertension, accounting for patient-level factors, socioeconomic context, and temporal dynamics.

Table 1 Descriptive statistics of selected variables

Variable	Variable description	Mean	SD
Female	Female (1 = female, 0 = male)	0.489	–
Age	Age of patient (years)	62.838	8.544
IMD1	IMD quintile 1 (least deprived)	0.265	–
IMD2	IMD quintile 2	0.231	–
IMD3	IMD quintile 3	0.196	–
IMD4	IMD quintile 4	0.165	–
IMD5	IMD quintile 5 (most deprived)	0.143	–
Number of diagnoses	Number of recorded diagnoses	2.019	1.764
ED	Emergency department visits (count)	0.247	0.711
GP	General practitioner visits (count)	4.960	4.836

Number of observations = 97,470; Number of patients = 16,245

Table 1 presents summary statistics for the key variables used in the analysis. The dataset includes 97,470 observations corresponding to 16,245 unique patients. The average age of patients is approximately 62.8 years (SD = 8.54), indicating an older population. Around 49% of the sample are female. Socioeconomic status is represented by the Index of Multiple Deprivation (IMD) quintiles, with IMD1 (least deprived) to IMD5 (most deprived) included as binary indicators; the proportion of patients in each quintile decreases from IMD1 (26.5%) to IMD5 (14.3%). Patients have, on average, just over two recorded diagnoses (mean = 2.019, SD = 1.764), suggesting a moderately complex clinical profile. The primary outcome variables are counts of healthcare visits: the average number of emergency department (ED) visits is 0.247 (SD = 0.711), whereas general practitioner (GP) visits are more frequent, with a mean of 4.960 (SD = 4.836). These statistics highlight variation in healthcare utilization across the sample and motivate the need for multivariate count modeling to account for overdispersion and correlation between ED and GP visits.

Figure 1 shows the distribution of within-practice Pearson correlations between GP and ED visit counts. The distribution is unimodal and centered slightly above zero, suggesting a potential association between the two outcomes across practices.

Figure 2 plots these correlations against the number of patients per practice. No systematic pattern emerges, although slightly more extreme values appear in smaller practices, likely due to higher estimation variability.

These findings indicate that the extent to which GP and ED use are correlated at the patient level varies across practices.

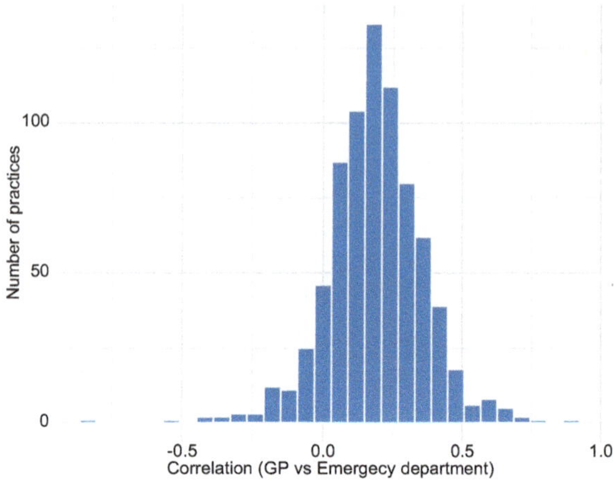

Fig. 1 Distribution of within-practice correlations between GP and ED visits

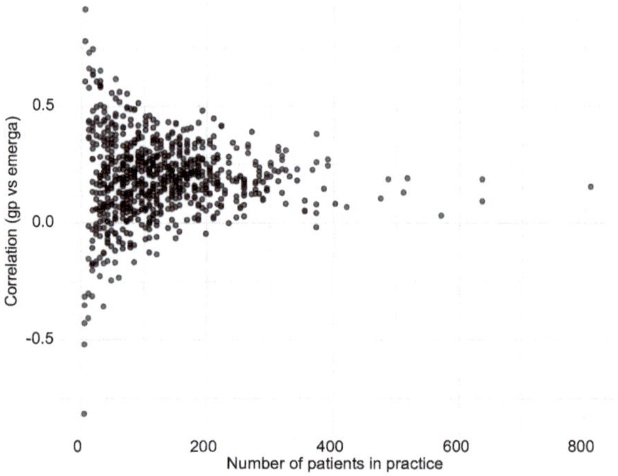

Fig. 2 Correlation between GP and ED visits within each practice, by practice size

3 The Model

We propose a Bayesian hierarchical bivariate Poisson model to jointly analyze emergency department (ED) visits and general practitioner (GP) consultations, accounting for their potential correlation while adjusting for individual- and practice-level effects. Let Y_{1it} and Y_{2it} denote, respectively, the observed number of emergency department (ED) visits and general practitioner (GP) consultations for individual i, who is enrolled in GP practice $g = 1, \ldots, G$ and observed at time $t = 1, \ldots, T_i$. These two count outcomes are modeled conditionally on covariates and latent effects via log-linear predictors. Specifically, we assume that the expected counts μ_{1it} and μ_{2it} follow the exponential link:

$$\mu_{1it} = \exp\left(\alpha_1 + \sum_{g=1}^{G} a_{1g} d_{igt} + b_{1i} + \mathbf{x}'_{1it}\boldsymbol{\beta}_1\right),$$

$$\mu_{2it} = \exp\left(\alpha_2 + \sum_{g=1}^{G} a_{2g} d_{igt} + b_{2i} + \mathbf{x}'_{2it}\boldsymbol{\beta}_2\right),$$

where α_1 and α_2 are intercepts, $\mathbf{x}_{1it}$ and $\mathbf{x}_{2it}$ are vectors of covariates, and $\boldsymbol{\beta}_1$, $\boldsymbol{\beta}_2$ are the corresponding regression coefficients. The terms b_{1i} and b_{2i} represent subject-specific random effects capturing individual heterogeneity. The variables d_{igt} are dummy indicators equal to one if individual i is enrolled in practice g at time t, and zero otherwise; the associated coefficients a_{1g} and a_{2g} are fixed effects accounting for systematic differences across GP practices in ED and GP utilization, respectively.

To model the marginal joint distribution of Y_{1it} and Y_{2it}, we employ a Bivariate Poisson distribution (Famoye, 2010; Lakshminarayana et al., 1999), Marginally, the two count variables are assumed to follow independent Poisson distributions:

$$Y_{1it} \sim \text{Po}(\mu_{1it}), \qquad Y_{2it} \sim \text{Po}(\mu_{2it}),$$

where Po is the Poisson distribution. However, to allow for statistical dependence between ED visits and GP consultations, we specify a bivariate likelihood in which the joint distribution of (Y_{1it}, Y_{2it}) is defined as a product of the two marginal Poisson likelihoods and a multiplicative correction term that introduces dependence. Specifically, we write

$$P(Y_{1it} = y_1, Y_{2it} = y_2) \propto \text{Po}(y_1 \mid \mu_{1it}) \cdot \text{Po}(y_2 \mid \mu_{2it}) \cdot A_{it}(y_1, y_2; \mu_{1it}, \mu_{2it}, \lambda),$$

where the adjustment function A_{it} is given by

$$A_{it} = 1 + \lambda(e^{-y_1} - e^{-q\mu_{1it}})(e^{-y_2} - e^{-q\mu_{2it}}),$$

with $\lambda \in [0, 1]$ controlling the strength of dependence and $q = 1 - e^{-1}$ ensuring proper scaling. This functional form implies that the dependence between the two outcomes increases with both λ and the proximity of the observed counts y_1, y_2 to their expected values.

To ensure model identifiability when including all G GP practice indicators, we impose a sum-to-zero constraint on the fixed effects associated with each practice. That is, we require that

$$\sum_{g=1}^{G} a_{1g} = 0, \quad \sum_{g=1}^{G} a_{2g} = 0,$$

so that practice-level effects are interpreted relative to the overall mean. Finally, individual-level heterogeneity is modeled via Gaussian random effects, where

$$b_{1i} \sim \mathcal{N}(0, \sigma_{1b}^2), \quad b_{2i} \sim \mathcal{N}(0, \sigma_{2b}^2).$$

These terms capture unobserved variation in the intensity of ED and GP use across patients, independently of observed covariates and GP affiliation.

4 Estimation

The estimation of the proposed model poses substantial computational challenges due to the presence of multiple layers of complexity. First, the specification includes random effects at the individual level, combined with a non-standard bivariate likelihood function that introduces nonlinear dependence between the two count outcomes with the dependence term A_{it}. Integrating over the high-dimensional individual-level random effects is already nontrivial in univariate Poisson models and becomes even more demanding in the bivariate case, where the likelihood does not factorize.

Second, the inclusion of fixed effects for a potentially large number of GP practices further increases the dimensionality of the parameter space. When the number of groups is high, classical methods such as maximum likelihood or pseudo-likelihood estimation may become unstable or computationally prohibitive, particularly one needs to account for uncertainty in both fixed and random components.

Moreover, traditional maximum likelihood estimation encounters three major difficulties when applied to hierarchical bivariate Poisson models. First, the bivariate Poisson likelihood involves a non-standard form due to the dependence parameter (λ), making analytical solutions intractable and requiring numerical approximations that may be unstable. Second, the presence of multiple layers of random effects (e.g., spatial, temporal, or subject-specific) introduces high-dimensional integration problems, where standard numerical methods (e.g., Laplace approximation or adaptive quadrature) become computationally prohibitive or inaccurate. Third, MLE provides only asymptotic approximations for uncertainty in random effects, often underestimating variance and leading to overly confident inferences, particularly in small-sample settings.

Therefore, we adopt a fully Bayesian approach to estimation, which allows for coherent uncertainty quantification and efficient integration over high-dimensional latent structures. The full posterior distribution is defined as follows:

$$p(\theta, \mathbf{a}_1, \mathbf{a}_2, \mathbf{b}_1, \mathbf{b}_2 \mid \mathbf{Y}_1, \mathbf{Y}_2, \mathbf{X}_1, \mathbf{X}_2) \propto \mathcal{L}(\mathbf{Y}_1, \mathbf{Y}_2 \mid \theta, \mathbf{a}_1, \mathbf{a}_2, \mathbf{b}_1, \mathbf{b}_2) \cdot p(\theta) \cdot p(\mathbf{a}_1) p(\mathbf{a}_2) p(\mathbf{b}_1) p(\mathbf{b}_2),$$

where $\theta = (\alpha_1, \alpha_2, \boldsymbol{\beta}_1, \boldsymbol{\beta}_2, \lambda)$.

We assign weakly informative priors to all model parameters. The intercepts α_1 and α_2, as well as the regression coefficients $\boldsymbol{\beta}_1$ and $\boldsymbol{\beta}_2$, follow Gaussian priors with mean 0 and variance 10^2. The dependence parameter λ is assigned a uniform prior on the unit interval, and the individual-level random effects are modeled as Gaussian with mean zero and variance parameters to be estimated: $b_{1i} \sim \mathcal{N}(0, \sigma^2_{1b})$ and $b_{2i} \sim \mathcal{N}(0, \sigma^2_{2b})$.

The individual log-likelihood contribution at time t is computed as the sum of the two Poisson log-probabilities and the dependence adjustment term:

$$\log \mathcal{L}_{it} = \log Po(Y_{1it} \mid \mu_{1it}) + \log Po(Y_{2it} \mid \mu_{2it}) + \log A_{it}.$$

To ensure identifiability of the GP fixed effects, we impose a sum-to-zero constraint across groups by reparameterization. In particular, the effect for the reference group (say, $g = G$) is defined as the negative sum of the effects for the remaining $G - 1$ groups:

$$a_{1G} = -\sum_{g=1}^{G-1} a_{1g}, \quad a_{2G} = -\sum_{g=1}^{G-1} a_{2g}.$$

The posterior distribution is sampled using Hamiltonian Monte Carlo (HMC), specifically the No-U-Turn Sampler (NUTS), as implemented in `Stan` and accessed via the `CmdStanR` interface in `R`. HMC introduces auxiliary momentum variables and simulates Hamiltonian dynamics in order to efficiently propose updates in the parameter space. The resulting trajectory follows the system:

$$\begin{aligned} \frac{d\theta}{dt} &= \frac{\partial H}{\partial r} = r, \\ \frac{dr}{dt} &= -\frac{\partial H}{\partial \theta} = \nabla_\theta \log p(\theta \mid data), \end{aligned}$$

where the Hamiltonian is defined as $H(\theta, r) = -\log p(\theta \mid data) + \frac{1}{2} r^\top r$. NUTS adaptively selects the integration path length by recursively building a binary tree and stopping when a U-turn is detected, thereby eliminating the need to tune the number of leapfrog steps. During warm-up, the algorithm also adapts the step size and the mass matrix to the local geometry of the posterior, improving sampling efficiency in complex hierarchical models.

We run four parallel chains with 2,000 warm-up iterations and 8,000 post-warm-up iterations per chain. Convergence is assessed through standard diagnostics, including the Gelman–Rubin statistic $\hat{R}$ (targeting values close to 1) and the effective sample size (ESS).

5 Results

Table 2 presents maximum likelihood estimates from pooled models of GP and Emergency Department (ED) visits. These models do not include GP fixed effects or individual random effects, as the likelihood becomes too complex and fails to converge. Moreover, maximum likelihood estimation with GP fixed effects alone also shows convergence issues, raising concerns about the model's stability when a large number of fixed effects are included.

Several key findings emerge. First, there is clear evidence of significant year effects (yd variables), particularly for later years. For instance, `Year 2013` and `Year 2014` are associated with increases in both GP and ED visits, though the effects are larger in the emergency model. This suggests increasing service use over time, especially for acute care.

Second, female patients are significantly more likely to visit GPs, with a coefficient of 0.074, while the effect is negligible and non-significant for emergency visits. Age has a small but positive and statistically significant effect in both models. Area-level deprivation effects (IMD) are more pronounced in the emergency model. Particularly, individuals in the most deprived group (`IMD5`) have a notably higher likelihood of ED use, with minimal differences across IMD groups in the GP model. Health status, captured by the total number of diagnoses, shows strong positive associations with both GP and emergency visits. The quadratic term is negative and significant, indicating a diminishing marginal effect as morbidity increases.

Lagged utilization variables also behave as expected. Previous GP use significantly predicts future GP visits, while prior emergency use strongly predicts future emergency visits, suggesting persistent patterns in care-seeking behavior. The intercepts indicate baseline utilization, with a much lower expected count for emergency visits (α = -2.317) compared to GP visits (0.735), consistent with the relative frequency of these types of care.

Overall, the model reveals distinct predictors and dynamics across GP and ED use, highlighting the importance of accounting for both individual and contextual factors in analyzing healthcare utilization (Table 2).

Table 3 reports the posterior estimates of covariate effects on ED visits. Temporal effects indicate a significant drop in visits in 2013, 2014, and 2015, followed by a sharp increase in 2016 and 2017, with the largest positive effect observed in 2017. Female patients show a slight positive association with ED visits (mean = 0.021). Age is positively associated with ED utilization (mean = 0.006, 95% CI: 0.003 to 0.008), indicating older individuals tend to have higher visit rates. Deprivation, captured via IMD quintiles, shows that more deprived groups (especially IMD5) are significantly more likely to attend ED; for example, the effect of being in IMD5 is 0.176 (95% CI: 0.087 to 0.264). The number of diagnoses has a strong positive linear effect (mean = 0.3), while the squared term has a negative effect, suggesting a concave relationship. Lagged ED visits (ED_{t-1}) have a substantial negative autoregressive coefficient (mean = -0.112), indicating that prior visits are associated with reduced future risk, possibly due to treatment or admission. In general, these results indicate

Table 2 ML estimates pooled

	GP		Emergency	
	Coef.	S.E.	Coef.	S.E.
Year 2013	0.013	0.005	−0.057	0.024
Year 2014	−0.026	0.005	−0.001	0.024
Year 2015	0.018	0.005	−0.014	0.024
Year 2016	0.033	0.005	0.139	0.023
Year 2017	0.015	0.005	0.215	0.022
Female	0.074	0.003	−0.003	0.013
Age	0.001	0.000	0.004	0.001
IMD2	0.008	0.004	0.085	0.019
IMD3	−0.003	0.004	0.113	0.020
IMD4	0.001	0.005	0.174	0.020
IMD5	0.034	0.005	0.219	0.021
Number of diagnoses	0.161	0.002	0.237	0.009
Number of diagnoses^2	−0.011	0.000	−0.011	0.001
ED_{t-1}	−0.003	0.002	0.252	0.007
GP_{t-1}	0.069	0.000	0.001	0.001
α	0.735	0.012	−2.317	0.053

the existence at the patient level of a substitution effect between GP and ED visits, with a relatively small persistence in ED utilization.

Table 4 presents the posterior estimates of covariate effects on GP visits. Temporal patterns show a decline in GP visits during the earlier years of the study, with statistically significant negative coefficients. However, coefficients for 2016 and 2017 are close to zero and not statistically significant, indicating a plateau in GP usage. Female patients are significantly more likely to use GP services than males, while age has a positive effect on GP visits, consistent with greater healthcare demand among older individuals. As expected, the number of diagnoses is strongly associated with more GP visits, while the squared term is negative, capturing nonlinear multimorbidity effects.

Lagged ED visits ED_{t-1}) have a small positive association, while lagged GP visits (GP_{t-1}) have a very strong positive effect (0.341), indicating persistence in GP usage over time.

Table 5 reports the posterior summaries of key structural parameters in the bivariate Poisson model. The parameter λ, which governs the degree of dependence between the two count outcomes, has a posterior mean of 0.959, suggesting strong and statistically significant positive dependence between the outcomes. The dispersion parameters σ_1 and σ_2, representing heterogeneity in the two marginal distributions, are estimated at 1.090 and 0.464, respectively. These values indicate moderate overdispersion for the first outcome and relatively low dispersion for the second.

Table 3 Covariate effects on ED visits

	Mean	SD	5%	95%
Year 2013	−0.074	0.024	−0.113	−0.035
Year 2014	−0.035	0.024	−0.073	0.005
Year 2015	−0.053	0.024	−0.092	−0.015
Year 2016	0.089	0.023	0.051	0.126
Year 2017	0.182	0.022	0.145	0.219
Female	0.021	0.025	−0.020	0.062
Age	0.006	0.002	0.003	0.008
IMD2	0.121	0.039	0.056	0.185
IMD3	0.095	0.043	0.024	0.167
IMD5	0.176	0.054	0.087	0.264
Number of diagnoses	0.300	0.014	0.277	0.324
Number of diagnoses^2	−0.012	0.002	−0.015	−0.008
ED_{t-1}	−0.112	0.008	−0.125	−0.100
GP_{t-1}	−0.004	0.002	−0.007	0.002
α_1	−3.020	0.100	−3.180	−2.850

Table 4 Covariate effects on GP visits

	Mean	SD	5%	95%
Year 2013	−0.091	0.005	−0.099	−0.083
Year 2014	−0.173	0.005	−0.182	−0.165
Year 2015	−0.162	0.005	−0.170	−0.153
Year 2016	−0.151	0.005	−0.159	−0.143
Year 2017	−0.168	0.005	−0.177	−0.160
Female	0.124	0.008	0.110	0.138
Age	0.002	0.001	0.001	0.003
IMD2	0.004	0.013	−0.017	0.026
IMD3	−0.016	0.014	−0.040	0.008
IMD4	−0.010	0.017	−0.038	0.018
IMD5	0.010	0.019	−0.021	0.041
Number of diagnoses	0.232	0.004	0.225	0.238
Number of diagnoses^2	−0.011	0.001	−0.011	−0.010
ED_{t-1}	−0.001	0.002	−0.004	0.003
GP_{t-1}	0.018	0.001	0.017	0.018
α_2	0.817	0.033	0.763	0.873

Table 5 Posterior summaries of selected structural parameters

Parameter	Mean	SD	5%	95%
λ	0.959	0.038	0.895	1.020
σ_1	1.090	0.014	1.070	1.120
σ_2	0.464	0.004	0.458	0.470

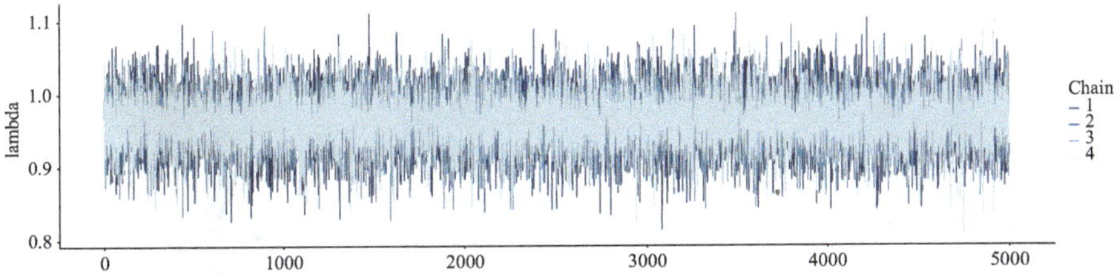

Fig. 3 Trace plot for the posterior distribution of λ across four chains

The narrow credible intervals suggest high precision in the posterior estimation of structural variability and dependence.

Figure 3 shows the trace plot for the posterior samples of the parameter λ across four independent Markov Chain Monte Carlo (MCMC) chains. Each line represents one chain, and the y-axis indicates the sampled values of λ over iterations.

The plot shows good mixing and convergence behavior. All four chains appear to be sampling from the same target distribution, with overlapping trajectories and no evidence of systematic divergence or trends. The chains rapidly explore the parameter space and fluctuate around a stable region, suggesting that the posterior distribution is being adequately explored.

Practice-level heterogeneity in ED and GP visit rates is depicted in Figs. 4 and 5. Particularly, the first figure shows the estimated random effects at the general practice level on the log rate of emergency department (ED) visits. Each point represents the posterior mean effect for a given practice, with horizontal lines indicating 90% credible intervals. Practices are ordered from lowest to highest effect. The spread of effects (ranging approximately from -3 to $+2.5$ on the log scale) highlights substantial heterogeneity across practices. Several intervals lie entirely above or below zero, indicating strong posterior evidence of deviation from the average ED usage.

Figure 5 shows analogous effects for general practitioner (GP) visits. Effects range from -4.5 to $+2$, with a concentration near zero or slightly above. This suggests that most practices provide at least average access to GP services, though some are markedly below the mean. The heterogeneity in both outcomes motivates a comparison of their joint distribution.

We now focus our attention on the relationship between ED and GP visit propensities. Figure 6 displays the joint distribution of the posterior means for the practice-level random effects on ED visits (a_1) and GP visits (a_2), both on the log-rate scale.

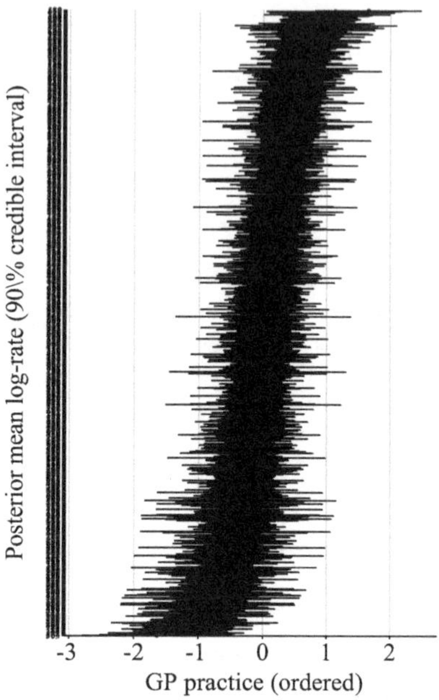

Fig. 4 Practice-level random effects on ED visits (a_1)

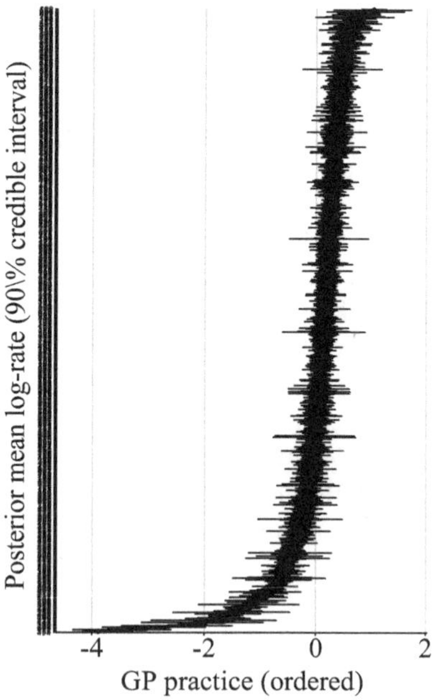

Fig. 5 Practice-level random effects on GP visits (a_2)

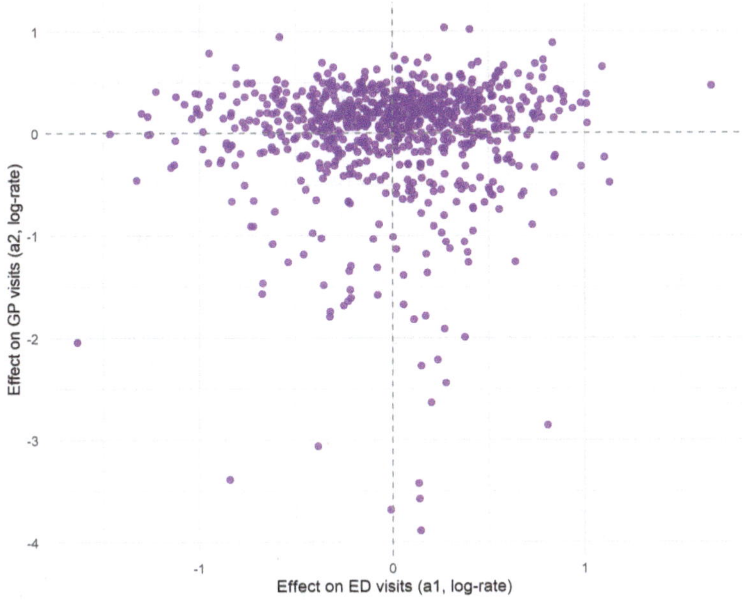

Fig. 6 Relationship between ED (a_{1g}) and GP (a_{2g}) effects by practice

Each point represents a GP practice. The cloud of points is centered around the origin, with no clear linear trend. The empirical correlation between the two effects is weak ($\rho \approx 0.15$), but statistically significant (p-value < 0.05), suggesting little systematic association between emergency and primary care at the practice level.

Our estimation findings align well with a growing literature on health-service count modeling. As observed in (Mohammadi et al., 2016), who modeled blood donation and deferral counts using a Bayesian BZIP model, incorporating both correlation and zero inflation can significantly improve predictive performance and interpretability (Mohammadi et al., 2016). Likewise, Kouakou et al. (2021) applied a zero-inflated bivariate Poisson framework to medical visit data and found better fit—based on DIC and predictive checks—than independent univariate models. These results resonate with our findings where both overdispersion and excess zeros are present, and dependency between service types (GP and ED visits) emerges strongly in the posterior correlation estimates.

Our use of hierarchical modeling to capture clustering at the patient and practice levels mirrors approaches in cost-effectiveness and healthcare utilization research. For instance, Grieve et al. (2010) extended Bayesian hierarchical models for cost–outcome pairs generated within cluster randomized trials to properly account for within-cluster correlation and heterogeneity, yielding more accurate uncertainty estimates. Additionally, spatial and hierarchical modeling techniques applied in small-area health-service studies—such as those linking spatial random effects with utilization rates—offer robust parallels to our own multilevel framework (MacNab, 2003).

The positive dependence between ED and GP visits observed in our structural parameter estimates likewise echoes findings from studies of co-occurring health-service use, such as asthma-related emergency room visits modeled via latent autoregressive Poisson processes (Al-Wahsh & Hussein, 2020), where correlated time-series counts required a bivariate Poisson with shared latent structure. Our results thus contribute to and extend a consistent pattern in the literature regarding joint health-service utilization modeling.

These references situate our results within methodological and applied contexts, underscoring the relevance of hierarchical bivariate Poisson models for accurately capturing dependence, overdispersion, zero inflation, and clustering in health outcome data.

6 Discussion and Conclusion

This study applies a Bayesian hierarchical bivariate Poisson framework to jointly analyze general practitioner (GP) consultations and emergency department (ED) visits among older adults newly diagnosed with hypertension, using rich longitudinal records from the UK CPRD database. By modeling correlated service outcomes and incorporating both individual- and practice-level factors, we link patterns of healthcare use to broader policy concerns around chronic disease management, system efficiency, and equity.

From a methodological point of view, by incorporating practice-level effects, the model highlights how much variation in utilization is attributable to patient characteristics (e.g., age, comorbidities, deprivation) versus organizational or contextual differences across providers. This distinction is directly relevant to policy: while individual-level variation may be addressed through targeted interventions, systematic differences across practices suggest the need for structural reforms, resource reallocation, or adjustments in commissioning and funding models.

Despite these advantages, several limitations must be acknowledged. The classical bivariate Poisson model assumes non-negative correlation and equidispersion in the marginal distributions, which may not fully reflect the realities of health-service use. In practice, excess zeros (e.g., patients with no ED visits), overdispersion, and even negative correlations may occur when services substitute for one another. In such cases, extensions such as zero-inflated, Poisson-lognormal, or "zero-dependent" models provide more accurate representations (Gonçalves et al., 2025; Kouakou et al., 2021; Qarmalah & Alzaid, 2023). Choosing between these alternatives should be guided not only by model fit criteria (e.g., DIC, WAIC) but also by substantive policy questions—such as whether the policy interest lies in identifying over-users, non-users, or substitution effects.

Another challenge is scalability. Applying hierarchical models to large administrative datasets, such as CPRD, can be computationally demanding. Nevertheless, these investments in computation are justified when the results inform resource allocation decisions that affect millions of patients. Where necessary, approximations or carefully designed hierarchical structures can ensure tractability without sacrificing policy relevance.

Our findings yield four main insights. First, multimorbidity emerges as a strong driver of both GP and ED utilization, underscoring the importance of integrated care pathways for patients with complex health needs. Second, socioeconomic deprivation is more strongly associated with ED attendance than with GP use, pointing to persistent inequalities in urgent care access even within a universal health system. Third, we observe meaningful heterogeneity across GP practices, indicating that organizational structures and local resources shape utilization, although their influence is modest compared to patient-level factors. Finally, the correlation between GP and ED use at the practice level is small but positive, suggesting a potential substitution between services. Although this effect is small in our sample of patients with hypertension, from a policy point of view, this provides additional evidence that better management at GP levels of patients with hypertension reduces additional ED visits for any cause.

Taken together, these findings suggest that expanding GP capacity on its own will not ease hospital pressure unless it is also accompanied by policies that also shift individuals' preferences—e.g., toward prevention and routine GP visits. Effective strategies must directly address patient-level drivers of demand—for example, supporting individuals with multimorbidity, improving chronic disease management, strengthening health literacy, and reducing barriers to timely access for disadvantaged groups.

Acknowledgement Catia Nicodemo received funding from the European Union. Views and opinions expressed are however those of the author(s) only and do not necessarily reflect those of the European Union or the granting authority European Union's Horizon Europe Research and Innovation program. Neither the European Union nor the granting authority can be held responsible for them. The PCR-4-ALL project has received funding under the Horizon Europe Research and Innovation program (the grant agreement No. 101095606). Catia Nicodemo has also received support from the National Institute for Health and Care Research (NIHR) School for Primary Care Research [Grant Reference Number: 588].

References

Al-Wahsh, H., & Hussein, A. (2020). A bivariate autoregressive poisson model and its application to asthma-related emergency room visits. *Statistics in Medicine, 39*, 3184–3194.

Berkhout, P., & Plug, E. (2004). A bivariate poisson count data model using conditional probabilities. *Statistica Neerlandica, 58*, 349–364.

Cameron, A. C., & Johansson, P. (1998). Count data models for health-care demand: A comparative study of functional form. *Journal of Applied Econometrics, 13*, 329–347.

Chib, S., & Winkelmann, R. (2001). Likelihood inference for model parameters in a multivariate poisson regression model. *Journal of Econometrics, 103*, 273–301.

Famoye, F. (2010). On the bivariate negative binomial regression model. *Journal of Applied Statistics, 37*, 969–981.

Gonçalves, J. N., Barreto-Souza, W., & Ombao, H. (2025). Paired count regressions for modeling the number of doctor consultations and non-prescribed drugs intake. *Statistical Methods in Medical Research, 34*, 1553–1573.

Grieve, R., Nixon, R., & Thompson, S. G. (2010). Bayesian hierarchical models for cost-effectiveness analyses that use data from cluster randomized trials. *Medical Decision Making, 30*, 427–434.

Herrett, E., Gallagher, A. M., Bhaskaran, K., Forbes, H., Mathur, R., Van Staa, T., & Smeeth, L. (2015). Data resource profile: clinical practice research datalink (cprd). *International Journal of Epidemiology, 44*, 827–836.

Karlis, D. (2006). Bayesian analysis of the differences of count data. *Statistics in Medicine, 25*, 2178–2195.

Khafrim, S., Kazemnejad, A., & Eskandari, F. (2008). Hierarchical bayesian analysis of bivariate poisson regression model. *World Applied Sciences Journal, 4*, 667–675.

Kouakou, K., et al. (2021). Zero-inflated bivariate poisson model: Application to healthcare demand data. *Statistical Methods in Medical Research, 30*, 255–270.

Kouakou, K., Hili, O., & Dupuy, J. (2021). Estimation in the zero-inflated bivariate poisson model with an application to health-care utilization data. *Afrika Statistika, 16*, 2767–2788.

Lakshminarayana, J., Pandit, S., & Rao, K. S. (1999). On a bivariate poisson distribution. *Communications in Statistics - Theory and Methods, 28*, 267–276.

MacNab, Y. C. (2003). Hierarchical bayesian spatial modelling of small-area health service outcome and utilization rates. *Statistics in Medicine, 22*, 1761–1773.

Mathur, R., Bhaskaran, K., Chaturvedi, N., Leon, D. A., vanStaa, T., Grundy, E., & Smeeth, L. (2014). Completeness and usability of ethnicity data in uk-based primary care and hospital databases. *Journal of Public Health, 36*, 684–692.

Mohammadi, A., Kheiri, S., & Sedehi, M. (2016). Analysis of blood transfusion data using bivariate zero-inflated poisson model: A bayesian approach. *Computational and Mathematical Methods in Medicine, 2016*, 7878325.

Mponda, P., & Kaombe, G. (2024). Modeling fertility and education jointly: A bivariate poisson regression approach in malawi. *BMC Public Health, 24*, 112.

Munkin, M. D., & Trivedi, P. K. (1999). Simulation-based estimation of multivariate mixed poisson regression models. *Econometrics Journal, 2*, 29–48.

Qarmalah, A., & Alzaid, A. (2023). A novel zero-dependent bivariate poisson model with application to healthcare data. *Mathematics, 11*, 812.

Sengupta, P., Rao, R. H., & Dey, D. K. (2016). Bivariate doubly inflated poisson models with applications. *Statistical Methodology, 28*, 14–30.

Whelton, P. K., Carey, R. M., Aronow, W. S., Casey, D. E., Collins, K. J., Himmelfarb, C. D., DePalma, S. M., Gidding, S., Jamerson, K. A., Jones, D. W., MacLaughlin, E. J., Muntner, P., Ovbiagele, B., Smith, S. C., Spencer, C. C., Stafford, R. S., Taler, S. J., Thomas, R. J., Williams, K. A.,... Wright, J. T. (2018). 2017 acc/aha/aapa/abc/acpm/ags/apha/ash/aspc/nma/pcna guideline for the prevention, detection, evaluation, and management of high blood pressure in adults. *Journal of the American College of Cardiology,71*, e127–e248.
World Health Organization. (2021). Hypertension fact sheet. https://www.who.int/news-room/fact-sheets/detail/hypertension. Accessed July 21, 2025.

Kolibri 360 GmbH
Ferdinand-Jühlke-Straße 7
99095 Erfurt, Deutschland
produktsicherheit@kolibri360.de